CAMBRIDGE LIBRARY COLLECTION

Books of enduring scholarly value

Darwin

Two hundred years after his birth and 150 years after the publication of 'On the Origin of Species', Charles Darwin and his theories are still the focus of worldwide attention. This series offers not only works by Darwin, but also the writings of his mentors in Cambridge and elsewhere, and a survey of the impassioned scientific, philosophical and theological debates sparked by his 'dangerous idea'.

Lay Sermons, Addresses, and Reviews

Thomas Henry Huxley (1825–95) became known as 'Darwin's bulldog' because of his forceful and energetic support for Darwin's theory, especially at the notorious British Association meeting in Oxford in 1860. In fact, Huxley had some reservations about aspects of the theory, especially the element of gradual, continuous progress, but in public he was unwavering in his allegiance, saying in a letter to Darwin 'As for your doctrines I am prepared to go to the Stake if requisite'. In his 1870 essay collection Lay Sermons, Addresses, and Reviews, of which the title alone was designed to provoke controversy, he offers a variety of his writings, many of which were originally talks given to a range of audiences from the Geological Society to the South London Working Men's College, and including his own review of On the Origin of Species in the Westminster Review and a typically passionate response to two other reviews less favourable to Darwin.

Lay Sermons, Addresses, and Reviews

Thomas Henry Huxley

CAMBRIDGE UNIVERSITY PRESS

Cambridge, New York, Melbourne, Madrid, Cape Town, Singapore,
São Paolo, Delhi, Dubai, Tokyo

Published in the United States of America by Cambridge University Press, New York

www.cambridge.org
Information on this title: www.cambridge.org/9781108001564

© in this compilation Cambridge University Press 2009

This edition first published 1870
This digitally printed version 2009

ISBN 978-1-108-00156-4 Paperback

This book reproduces the text of the original edition. The content and language reflect
the beliefs, practices and terminology of their time, and have not been updated.

Cambridge University Press wishes to make clear that the book, unless originally published
by Cambridge, is not being republished by, in association or collaboration with, or
with the endorsement or approval of, the original publisher or its successors in title.

LAY SERMONS, ADDRESSES, AND REVIEWS.

LAY SERMONS, ADDRESSES,

AND

REVIEWS.

BY

THOMAS HENRY HUXLEY, LL.D., F.R.S.

London:

MACMILLAN AND CO.

1870.

A PREFATORY LETTER.

My dear Tyndall,

I should have liked to provide this collection of "Lay Sermons, Addresses, and Reviews," with a Dedication and a Preface. In the former, I should have asked you to allow me to associate your name with the book, chiefly on the ground that the oldest of the papers in it is a good deal younger than our friendship. In the latter, I intended to comment upon certain criticisms with which some of these Essays have been met.

But, on turning the matter over in my mind, I began to fear that a formal dedication at the beginning of such a volume would look like a grand lodge in front of a set of cottages ; while a complete defence of any of my old papers would simply amount to writing a new one—a labour for which I am, at present, by no means fit.

The book must go forth, therefore, without any better substitute for either Dedication, or Preface, than this letter ; before concluding which it is necessary for me to notify you, and any other reader, of two or three matters.

The first is, that the oldest Essay of the whole, that "On the Educational Value of the Natural History Sciences," contains a view of the nature of the differences between living and not-living bodies out of which I have long since grown.

Secondly, in the same paper, there is a statement concerning the method of the mathematical sciences, which, repeated and expanded elsewhere, brought upon me, during the meeting of the British Association at Exeter, the artillery of our eminent friend Professor Sylvester.

No one knows better than you do, how readily I should defer to the opinion of so great a mathematician if the question at issue were really, as he seems to think it is, a mathematical one. But I submit, that the dictum of a mathematical athlete upon a difficult problem which mathematics offers to philosophy, has no more special weight, than the verdict of that great pedestrian Captain Barclay would have had, in settling a disputed point in the physiology of locomotion.

The genius which sighs for new worlds to conquer beyond that surprising region in which "geometry, algebra, and the theory of numbers melt into one another like sunset tints, or the colours of a dying dolphin," may be of comparatively little service in the cold domain (mostly lighted by the moon, some say) of philosophy. And the more I think of it, the more does our friend seem to me to fall into the position of one of those "verständige Leute," about whom he makes so apt a quotation from Goethe. Surely he has not duly considered two points. The first, that I am in no way answerable for the origination of the doctrine he criticises : and the second, that if we are to employ the terms observation, induction, and experiment, in the sense in which he uses them, logic is as much an observational, inductive, and experimental science as mathematics; and that, I confess, appears to me to be a *reductio ad absurdum* of his argument.

Thirdly, the essay " On the Physical Basis of Life" was intended to contain a plain and untechnical statement of one of the great tendencies of modern biological thought, accompanied by a protest, from the philosophical side, against what is commonly called Materialism. The result of my well-meant efforts I find to be, that I am generally credited with having invented " protoplasm " in the interests of "materialism." My unlucky " Lay Sermon " has been attacked by microscopists, ignorant alike of Biology and Philosophy ; by philosophers, not very learned in either Biology or Microscopy ; by clergymen of several denominations ; and by some few writers who have taken the trouble to understand the subject. I trust that these last will believe that I leave the essay unaltered from no want of respectful attention to all they have said.

Fourthly, I wish to refer all who are interested in the topics discussed in my address on " Geological Reform," to the reply with which Sir William Thomson has honoured me.

And, lastly, let me say that I reprint the review of " The Origin of Species " simply because it has been cited as mine by a late President of the Geological Society.

If you find its phraseology, in some places, to be more vigorous than seems needful, recollect that it was written in the heat of our first battles over the Novum Organon of biology; that we were all ten years younger in those days; and last, but not least, that it was not published until it had been submitted to the revision of a friend for whose judgment I had then, as I have now, the greatest respect.

Ever, my dear TYNDALL,

Yours very faithfully,

T. H. HUXLEY

LONDON, _June_ 1870.

CONTENTS.

LAY SERMONS, ADDRESSES, AND REVIEWS.

I.

ON THE ADVISABLENESS OF
IMPROVING NATURAL KNOWLEDGE.

THIS time two hundred years ago—in the beginning of January, 1666—those of our forefathers who inhabited this great and ancient city, took breath between the shocks of two fearful calamities, one not quite past, although its fury had abated; the other to come.

Within a few yards of the very spot on which we are assembled, so the tradition runs, that painful and deadly malady, the plague, appeared in the latter months of 1664; and, though no new visitor, smote the people of England, and especially of her capital, with a violence unknown before, in the course of the following year. The hand of a master has pictured what happened in those dismal months; and in that truest of fictions, "The History of the Plague Year," Defoe shows death, with every accompaniment of pain and terror, stalking through the narrow streets of old London, and changing their busy hum into a silence broken only by the wailing of the mourners of fifty thousand dead; by the woful denunciations and mad prayers of fanatics; and by the madder yells of despairing profligates.

But, about this time in 1666, the death-rate had sunk to nearly its ordinary amount; a case of plague occurred only here and there, and the richer citizens who had flown from the pest had returned to their dwellings. The remnant of the people began to toil at the accustomed round of duty, or of pleasure; and the stream of city life bid fair to flow back along its old bed, with renewed and uninterrupted vigour.

The newly kindled hope was deceitful. The great plague, indeed, returned no more; but what it had done for the Londoners, the great fire, which broke out in the autumn of 1666, did for London; and, in September of that year, a heap of ashes and the indestructible energy of the people were all that remained of the glory of five-sixths of the city within the walls.

Our forefathers had their own ways of accounting for each of these calamities. They submitted to the plague in humility and in penitence, for they believed it to be the judgment of God. But, towards the fire they were furiously indignant, interpreting it as the effect of the malice of man,—as the work of the Republicans, or of the Papists, according as their prepossessions ran in favour of loyalty or of Puritanism.

It would, I fancy, have fared but ill with one who, standing where I now stand, in what was then a thickly peopled and fashionable part of London, should have broached to our ancestors the doctrine which I now propound to you—that all their hypotheses were alike wrong; that the plague was no more, in their sense, Divine judgment, than the fire was the work of any political, or of any religious, sect; but that they were themselves the authors of both plague and fire, and that they must look to themselves to prevent the recurrence of calamities, to all appearance so peculiarly beyond the reach of human control—so evidently the result of the wrath of God, or of the craft and subtlety of an enemy.

And one may picture to oneself how harmoniously the holy cursing of the Puritan of that day would have chimed in with the unholy cursing and the crackling wit of the Rochesters and Sedleys, and with the revilings of the political fanatics, if my imaginary plain dealer had gone on to say that, if the return of such misfortunes were ever rendered impossible, it would not be in virtue of the victory of the faith of Laud, or of that of Milton; and, as little, by the triumph of republicanism, as by that of monarchy. But that the one thing needful for compassing this end was, that the people of England should second the efforts of an insignificant corporation, the establishment of which, a few years before the epoch of the great plague and the great fire, had been as little noticed, as they were conspicuous.

Some twenty years before the outbreak of the plague a few calm and thoughtful students banded themselves together for the purpose, as they phrased it, of "improving natural knowledge." The ends they proposed to attain cannot be stated more clearly than in the words of one of the founders of the organization :—

" Our business was (precluding matters of theology and state affairs) to discourse and consider of philosophical enquiries, and such as related thereunto :—as Physick, Anatomy, Geometry, Astronomy, Navigation, Staticks, Magneticks, Chymicks, Mechanicks, and

Natural Experiments; with the state of these studies and their cultivation at home and abroad. We then discoursed of the circulation of the blood, the valves in the veins, the venæ lacteæ, the lymphatic vessels, the Copernican hypothesis, the nature of comets and new stars, the satellites of Jupiter, the oval shape (as it then appeared) of Saturn, the spots on the sun and its turning on its own axis, the inequalities and seleno-graphy of the moon, the several phases of Venus and Mercury, the improvement of telescopes and grinding of glasses for that purpose, the weight of air, the possibility or impossibility of vacuities and nature's ab-horrence thereof, the Torricellian experiment in quick-silver, the descent of heavy bodies and the degree of acceleration therein, with divers other things of like nature, some of which were then but new discoveries, and others not so generally known and embraced as now they are; with other things appertaining to what hath been called the New Philosophy, which, from the times of Galileo at Florence, and Sir Francis Bacon (Lord Verulam) in England, hath been much cultivated in Italy, France, Germany, and other parts abroad, as well as with us in England."

The learned Dr. Wallis, writing in 1696, narrates, in these words, what happened half a century before, or about 1645. The associates met at Oxford, in the rooms of Dr. Wilkins, who was destined to become a bishop; and subsequently coming together in London, they attracted the notice of the king. And it is a strange evidence of the taste for knowledge which the most obviously worthless of the Stuarts shared with his father and grandfather, that Charles the Second was not content with saying witty things about his philosophers, but did wise things with regard to them. For he not only bestowed upon them such attention as he could spare from his poodles and his mistresses, but, being in his usual state of impecuniosity, begged for them of the Duke of Ormond ; and, that step being without effect, gave them Chelsea College, a charter, and a mace : crowning his favours in the best way they could be crowned, by burdening them no further with royal patronage or state interference.

Thus it was that the half-dozen young men, studious of the " New Philosophy," who met in one another's lodgings in Oxford or in London, in the middle of the seventeenth century, grew in numerical and in real strength, until, in its latter part, the " Royal Society for the Improvement of Natural Knowledge " had already become famous, and had acquired a claim upon the veneration of Englishmen, which it has ever since retained, as the principal focus of scientific activity in our islands, and the chief champion of the cause it was formed to support.

It was by the aid of the Royal Society that Newton published his " Principia." If all the books in the world, except the Philosophical Transactions, were destroyed, it is safe to say that the foundations of physical science would remain unshaken, and that the vast intellectual progress of the last two centuries would be largely, though incompletely, recorded. Nor have any signs of halting or of decrepitude manifested themselves in our own times. As in Dr. Wallis's days, so in these, " our business is, precluding theology and state affairs, to discourse and consider of philosophical enquiries."

But our "Mathematick" is one which Newton would have to go to school to learn ; our "Staticks, Mechanicks, Magneticks, Chymicks, and Natural Experiments" constitute a mass of physical and chemical knowledge, a glimpse at which would compensate Galileo for the doings of a score of inquisitorial cardinals ; our "Physick" and "Anatomy" have embraced such infinite varieties of being, have laid open such new worlds in time and space, have grappled, not unsuccessfully, with such complex problems, that the eyes of Vesalius and of Harvey might be dazzled by the sight of the tree that has grown out of their grain of mustard seed.

The fact is perhaps rather too much, than too little, forced upon one's notice, nowadays, that all this marvellous intellectual growth has a no less wonderful expression in practical life ; and that, in this respect, if in no other, the movement symbolized by the progress of the Royal Society stands without a parallel in the history of mankind.

A series of volumes as bulky as the Transactions of the Royal Society might possibly be filled with the subtle speculations of the schoolmen ; not improbably, the obtaining a mastery over the products of mediæval thought might necessitate an even greater expenditure of time and of energy than the acquirement of the "New Philosophy ;" but though such work engrossed the best intellects of Europe for a longer time than has elapsed since the great fire, its effects were "writ in water," so far as our social state is concerned.

On the other hand, if the noble first President of the Royal Society could revisit the upper air and once more gladden his eyes with a sight of the familiar mace, he would find himself in the midst of a material civilization more different from that of his day, than that of the seventeenth, was from that of the first, century. And if Lord Brouncker's native sagacity had not deserted his ghost, he would need no long reflection to discover that all these great ships, these railways, these telegraphs, these factories, these printing presses, without which the whole fabric of modern English society would collapse into a mass of stagnant and starving pauperism,—that all these pillars of our State are but the ripples and the bubbles upon the surface of that great spiritual stream, the springs of which, only, he and his fellows were privileged to see ; and seeing, to recognise as that which it behoved them above all things to keep pure and undefiled.

It may not be too great a flight of imagination to conceive our noble *revenant* not forgetful of the great troubles of his own day, and anxious to know how often London had been burned down since his time, and how often the plague had carried off its thousands. He would have to learn that, although London contains tenfold the inflammable matter that it did in 1666 ; though, not content with filling our rooms with woodwork and light draperies, we must needs lead inflammable and explosive gases into every corner of our streets and houses, we never allow even a street to burn down. And if he asked how this had come about, we should have to explain that the improvement of natural knowledge has furnished us with dozens of machines for throwing water upon fires, any one of which would have furnished the ingenious Mr. Hooke, the first " curator and ex-

perimenter" of the Royal Society, with ample materials
for discourse before half a dozen meetings of that body;
and that, to say truth, except for the progress of natural
knowledge, we should not have been able to make even
the tools by which these machines are constructed.
And, further, it would be necessary to add, that although
severe fires sometimes occur and inflict great damage,
the loss is very generally compensated by societies, the
operations of which have been rendered possible only
by the progress of natural knowledge in the direction of
mathematics, and the accumulation of wealth in virtue
of other natural knowledge.

But the plague? My Lord Brouncker's observation
would not, I fear, lead him to think that Englishmen of
the nineteenth century are purer in life, or more fervent in religious faith, than the generation which could
produce a Boyle, an Evelyn, and a Milton. He might
find the mud of society at the bottom, instead of at the
top, but I fear that the sum total would be as deserving
of swift judgment as at the time of the Restoration.
And it would be our duty to explain once more, and
this time not without shame, that we have no reason
to believe that it is the improvement of our faith, nor
that of our morals, which keeps the plague from our
city; but, again, that it is the improvement of our
natural knowledge.

We have learned that pestilences will only take up
their abode among those who have prepared unswept
and ungarnished residences for them. Their cities must
have narrow, unwatered streets, foul with accumulated
garbage. Their houses must be ill-drained, ill-lighted,
ill-ventilated. Their subjects must be ill-washed, ill-

fed, ill-clothed. The London of 1665 was such a city. The cities of the East, where plague has an enduring dwelling, are such cities. We, in later times, have learned somewhat of Nature, and partly obey her. Because of this partial improvement of our natural knowledge and of that fractional obedience, we have no plague ; because that knowledge is still very imperfect and that obedience yet incomplete, typhus is our companion and cholera our visitor. But it is not presumptuous to express the belief that, when our knowledge is more complete and our obedience the expression of our knowledge, London will count her centuries of freedom from typhus and cholera, as she now gratefully reckons her two hundred years of ignorance of that plague which swooped upon her thrice in the first half of the seventeenth century.

Surely, there is nothing in these explanations which is not fully borne out by the facts ? Surely, the principles involved in them are now admitted among the fixed beliefs of all thinking men ? Surely, it is true that our countrymen are less subject to fire, famine, pestilence, and all the evils which result from a want of command over and due anticipation of the course of Nature, than were the countrymen of Milton; and health, wealth, and well-being are more abundant with us than with them ? But no less certainly is the difference due to the improvement of our knowledge of Nature, and the extent to which that improved knowledge has been incorporated with the household words of men, and has supplied the springs of their daily actions.

Granting for a moment, then, the truth of that which the depreciators of natural knowledge are so fond of urging, that its improvement can only add to the resources of our material civilization; admitting it to be possible that the founders of the Royal Society themselves looked for no other reward than this, I cannot confess that I was guilty of exaggeration when I hinted, that to him who had the gift of distinguishing between prominent events and important events, the origin of a combined effort on the part of mankind to improve natural knowledge might have loomed larger than the Plague and have outshone the glare of the Fire; as a something fraught with a wealth of beneficence to mankind, in comparison with which the damage done by those ghastly evils would shrink into insignificance.

It is very certain that for every victim slain by the plague, hundreds of mankind exist and find a fair share of happiness in the world, by the aid of the spinning jenny. And the great fire, at its worst, could not have burned the supply of coal, the daily working of which, in the bowels of the earth, made possible by the steam pump, gives rise to an amount of wealth to which the millions lost in old London are but as an old song.

But spinning jenny and steam pump are, after all, but toys, possessing an accidental value; and natural knowledge creates multitudes of more subtle contrivances, the praises of which do not happen to be sung because they are not directly convertible into instruments for creating wealth. When I contemplate natural knowledge squandering such gifts among men, the only appropriate comparison I can find for her is, to liken her to such a peasant woman as one sees in the Alps, striding ever upward, heavily burdened, and with mind bent only on her home ; but yet, without effort and without thought, knitting for her children. Now stockings are good and comfortable things, and the children will undoubtedly be much the better for them ; but surely it would be short-sighted, to say the least of it, to depreciate this toiling mother as a mere stocking-machine—a mere provider of physical comforts ?

However, there are blind leaders of the blind, and not a few of them, who take this view of natural knowledge, and can see nothing in the bountiful mother of humanity but a sort of comfort-grinding machine. According to them, the improvement of natural knowledge always has been, and always must be, synonymous with no more than the improvement of the material resources and the increase of the gratifications of men.

Natural knowledge is, in their eyes, no real mother of mankind, bringing them up with kindness, and, if need be, with sternness, in the way they should go, and instructing them in all things needful for their welfare ; but a sort of fairy godmother, ready to furnish her pets with shoes of swiftness, swords of sharpness, and omnipotent Aladdin's lamps, so that they may have telegraphs to Saturn, and see the other side of the moon, and thank God they are better than their benighted ancestors.

If this talk were true, I, for one, should not greatly care to toil in the service of natural knowledge. I think I would just as soon be quietly chipping my own flint axe, after the manner of my forefathers a few thousand years back, as be troubled with the endless malady of thought which now infests us all, for such reward. But I venture to say that such views are contrary alike to reason and to fact. Those who discourse in such fashion seem to me to be so intent upon trying to see what is above Nature, or what is behind her, that they are blind to what stares them in the face, in her.

I should not venture to speak thus strongly if my justification were not to be found in the simplest and most obvious facts,—if it needed more than an appeal to the most notorious truths to justify my assertion, that the improvement of natural knowledge, whatever direction it has taken, and however low the aims of those who may have commenced it—has not only conferred practical benefits on men, but, in so doing, has effected a revolution in their conceptions of the universe and of themselves, and has profoundly altered their modes of thinking and their views of right and wrong. I say that natural knowledge, seeking to satisfy natural wants, has found the ideas which can alone still spiritual cravings. I say that natural knowledge, in desiring to ascertain the laws of comfort, has been driven to discover those of conduct, and to lay the foundations of a new morality.

Let us take these points separately; and, first, what great ideas has natural knowledge introduced into men's minds?

I cannot but think that the foundations of all natural knowledge were laid when the reason of man first came face to face with the facts of Nature: when the savage first learned that the fingers of one hand are fewer than those of both; that it is shorter to cross a stream than to head it; that a stone stops where it is unless it be moved, and that it drops from the hand which lets it go;

that light and heat come and go with the sun ; that sticks burn away in a fire ; that plants and animals grow and die ; that if he struck his fellow-savage a blow he would make him angry, and perhaps get a blow in return, while if he offered him a fruit he would please him, and perhaps receive a fish in exchange. When men had acquired this much knowledge, the outlines, rude though they were, of mathematics, of physics, of chemistry, of biology, of moral, economical, and political science, were sketched. Nor did the germ of religion fail when science began to bud. Listen to words which, though new, are yet three thousand years old :—

> " . . . When in heaven the stars about the moon
> Look beautiful, when all the winds are laid,
> And every height comes out, and jutting peak
> And valley, and the immeasurable heavens
> Break open to their highest, and all the stars
> Shine, and the shepherd gladdens in his heart." [1]

If the half-savage Greek could share our feelings thus far, it is irrational to doubt that he went further, to find, as we do, that upon that brief gladness there follows a certain sorrow,—the little light of awakened human intelligence shines so mere a spark amidst the abyss of the unknown and unknowable ; seems so insufficient to do more than illuminate the imperfections that cannot be remedied, the aspirations that cannot be realized, of man's own nature. But in this sadness, this consciousness of the limitation of man, this sense of an open secret which he cannot penetrate, lies the essence of all religion ; and the attempt to embody it in the forms furnished by the intellect is the origin of the higher theologies.

[1] Need it be said that this is Tennyson's English for Homer's Greek ?

Thus it seems impossible to imagine but that the foundations of all knowledge—secular or sacred—were laid when intelligence dawned, though the superstructure remained for long ages so slight and feeble as to be compatible with the existence of almost any general view respecting the mode of governance of the universe. No doubt, from the first, there were certain phenomena which, to the rudest mind, presented a constancy of occurrence, and suggested that a fixed order ruled, at any rate, among them. I doubt if the grossest of Fetish worshippers ever imagined that a stone must have a god within it to make it fall, or that a fruit had a god within it to make it taste sweet. With regard to such matters as these, it is hardly questionable that mankind from the first took strictly positive and scientific views.

But, with respect to all the less familiar occurrences which present themselves, uncultured man, no doubt, has always taken himself as the standard of comparison, as the centre and measure of the world ; nor could he well avoid doing so. And finding that his apparently un-caused will has a powerful effect in giving rise to many occurrences, he naturally enough ascribed other and greater events to other and greater volitions, and came to look upon the world and all that therein is, as the product of the volitions of persons like himself, but stronger, and capable of being appeased or angered, as he himself might be soothed or irritated. Through such conceptions of the plan and working of the universe all mankind have passed, or are passing. And we may now consider, what has been the effect of the improvement of natural knowledge on the views of men who have reached this stage, and who have begun to cultivate natural knowledge with no desire but that of "increasing God's honour and bettering man's estate."

For example: what could seem wiser, from a mere material point of view, more innocent, from a theological one, to an ancient people, than that they should learn the exact succession of the seasons, as warnings for their husbandmen; or the position of the stars, as guides to their rude navigators? But what has grown out of this search for natural knowledge of so merely useful a character? You all know the reply. Astronomy,—which of all sciences has filled men's minds with general ideas of a character most foreign to their daily experience, and has, more than any other, rendered it impossible for them to accept the beliefs of their fathers. Astronomy,—which tells them that this so vast and seemingly solid earth is but an atom among atoms, whirling, no man knows whither, through illimitable space; which demonstrates that what we call the peaceful heaven above us, is but that space, filled by an infinitely subtle matter whose particles are seething and surging, like the waves of an angry sea; which opens up to us infinite regions where nothing is known, or ever seems to have been known, but matter and force, operating according to rigid rules; which leads us to contemplate phenomena the very nature of which demonstrates that they must have had a beginning, and that they must have an end, but the very nature of which also proves that the beginning was, to our conceptions of time, infinitely remote, and that the end is as immeasurably distant.

But it is not alone those who pursue astronomy who

ask for bread and receive ideas. What more harmless than the attempt to lift and distribute water by pumping it; what more absolutely and grossly utilitarian? But out of pumps grew the discussions about Nature's abhorrence of a vacuum; and then it was discovered that Nature does not abhor a vacuum, but that air has weight; and that notion paved the way for the doctrine that all matter has weight, and that the force which produces weight is co-extensive with the universe,—in short, to the theory of universal gravitation and endless force. While learning how to handle gases led to the discovery of oxygen, and to modern chemistry, and to the notion of the indestructibility of matter.

Again, what simpler, or more absolutely practical, than the attempt to keep the axle of a wheel from heating when the wheel turns round very fast? How useful for carters and gig drivers to know something about this; and how good were it, if any ingenious person would find out the cause of such phenomena, and thence educe a general remedy for them. Such an ingenious person was Count Rumford; and he and his successors have landed us in the theory of the persistence, or indestructibility, of force. And in the infinitely minute, as in the infinitely great, the seekers after natural knowledge, of the kinds called physical and chemical, have everywhere found a definite order and succession of events which seem never to be infringed.

And how has it fared with "Physick" and Anatomy? Have the anatomist, the physiologist, or the physician, whose business it has been to devote themselves assiduously to that eminently practical and direct end, the alleviation of the sufferings of mankind,—have they been able to confine their vision more absolutely to the strictly useful? I fear they are worst offenders of all. For if the astronomer has set before us the infinite magnitude of space, and the practical eternity of the duration of the universe; if the physical and chemical philosophers have demonstrated the infinite minuteness of its constituent parts, and the practical eternity of matter and of force; and if both have alike proclaimed the universality of a definite and predicable order and succession of events, the workers in biology have not only accepted all these, but have added more startling theses of their own. For, as the astronomers discover in the earth no centre of the universe, but an eccentric speck, so the naturalists find man to be no centre of the living world, but one amidst endless modifications of life; and as the astronomer observes the mark of practically endless time set upon the arrangements of the solar system, so the student of life finds the records of ancient forms of existence peopling the world for ages, which, in relation to human experience, are infinite.

Furthermore, the physiologist finds life to be as dependent for its manifestation on particular molecular arrangements as any physical or chemical phenomenon; and, wherever he extends his researches, fixed order and unchanging causation reveal themselves, as plainly as in the rest of Nature.

Nor can I find that any other fate has awaited the germ of Religion. Arising, like all other kinds of knowledge, out of the action and interaction of man's mind, with that which is not man's mind, it has taken the intellectual coverings of Fetishism or Polytheism; of Theism or Atheism; of Superstition or Rationalism.

With these, and their relative merits and demerits, I have nothing to do; but this it is needful for my purpose to say, that if the religion of the present differs from that of the past, it is because the theology of the present has become more scientific than that of the past; because it has not only renounced idols of wood and idols of stone, but begins to see the necessity of breaking in pieces the idols built up of books and traditions and fine-spun ecclesiastical cobwebs : and of cherishing the noblest and most human of man's emotions, by worship " for the most part of the silent sort" at the altar of the Unknown and Unknowable.

Such are a few of the new conceptions implanted in our minds by the improvement of natural knowledge. Men have acquired the ideas of the practically infinite extent of the universe and of its practical eternity ; they are familiar with the conception that our earth is but an infinitesimal fragment of that part of the universe which can be seen ; and that, nevertheless, its duration is, as compared with our standards of time, infinite.   They have further acquired the idea that man is but one of innumerable forms of life now existing in the globe, and that the present existences are but the last of an immeasurable series of predecessors.   Moreover, every step they have made in natural knowledge has tended to extend and rivet in their minds the conception of a definite order of the universe—which is embodied in what are called, by an unhappy metaphor, the laws of Nature—and to narrow the range and loosen the force of men's belief in spontaneity, or in changes other than such as arise out of that definite order itself.

Whether these ideas are well or ill founded is not the question. No one can deny that they exist, and have been the inevitable outgrowth of the improvement of natural knowledge. And if so, it cannot be doubted that they are changing the form of men's most cherished and most important convictions.

And as regards the second point—the extent to which the improvement of natural knowledge has remodelled and altered what may. be termed the intellectual ethics of men,—what are among the moral convictions most fondly held by barbarous and semi-barbarous people?

They are the convictions that authority is the soundest basis of belief; that merit attaches to a readiness to believe; that the doubting disposition is a bad one, and scepticism a sin; that when good authority has pronounced what is to be believed, and faith has accepted it, reason has no further duty. There are many excellent persons who yet hold by these principles, and it is not my present business, or intention, to discuss their views. All I wish to bring clearly before your minds is the unquestionable fact, that the improvement of natural knowledge is effected by methods which directly give the lie to all these convictions, and assume the exact reverse of each to be true.

The improver of natural knowledge absolutely refuses to acknowledge authority, as such. For him, scepticism is the highest of duties; blind faith the one unpardonable sin. And it cannot be otherwise, for every great advance in natural knowledge has involved the absolute rejection of authority, the cherishing of the keenest scepticism, the annihilation of the spirit of blind faith:

and the most ardent votary of science holds his firmest convictions, not because the men he most venerates hold them; not because their verity is testified by portents and wonders; but because his experience teaches him that whenever he chooses to bring these convictions into contact with their primary source, Nature—whenever he thinks fit to test them by appealing to experiment and to observation—Nature will confirm them. The man of science has learned to believe in justification, not by faith, but by verification.

Thus, without for a moment pretending to despise the practical results of the improvement of natural knowledge, and its beneficial influence on material civilization, it must, I think, be admitted that the great ideas, some of which I have indicated, and the ethical spirit which I have endeavoured to sketch, in the few moments which remained at my disposal, constitute the real and permanent significance of natural knowledge.

If these ideas be destined, as I believe they are, to be more and more firmly established as the world grows older; if that spirit be fated, as I believe it is, to extend itself into all departments of human thought, and to become co-extensive with the range of knowledge; if, as our race approaches its maturity, it discovers, as I believe it will, that there is but one kind of knowledge and but one method of acquiring it; then we, who are still children, may justly feel it our highest duty to recognise the advisableness of improving natural knowledge, and so to aid ourselves and our successors in their course towards the noble goal which lies before mankind.

II.

EMANCIPATION—BLACK AND WHITE.

QUASHIE'S plaintive inquiry, "Am I not a man and a brother?" seems at last to have received its final reply— the recent decision of the fierce trial by battle on the other side of the Atlantic fully concurring with that long since delivered here in a more peaceful way.

The question is settled; but even those who are most thoroughly convinced that the doom is just, must see good grounds for repudiating half the arguments which have been employed by the winning side; and for doubting whether its ultimate results will embody the hopes of the victors, though they may more than realize the fears of the vanquished. It may be quite true that some negroes are better than some white men; but no rational man, cognizant of the facts, believes that the average negro is the equal, still less the superior, of the average white man. And, if this be true, it is simply incredible that, when all his disabilities are removed, and our prognathous relative has a fair field and no favour, as well as no oppressor, he will be able to compete successfully with his bigger-brained and smaller-jawed rival, in a contest which is to be carried on by thoughts and not by bites. The highest places in the hierarchy of civilization will assuredly not be within the reach of our dusky cousins, though it is by no means necessary that they should be restricted to the lowest. But whatever the position of stable equilibrium into which the laws of social gravitation may bring the negro, all responsibility for the result will henceforward lie between Nature and him. The white man may wash his hands of it, and the Caucasian conscience be void of reproach for evermore. And this, if we look to the bottom of the matter, is the real justification for the abolition policy.

The doctrine of equal natural rights may be an illogical delusion; emancipation may convert the slave from a well fed animal into a pauperised man; mankind may even have to do without cotton shirts; but all these evils must be faced, if the moral law, that no human being can arbitrarily dominate over another without grievous damage to his own nature, be, as many think, as readily demonstrable by experiment as any physical truth. If this be true, no slavery can be abolished without a double emancipation, and the master will benefit by freedom more than the freed-man.

The like considerations apply to all the other questions of emancipation which are at present stirring the world— the multifarious demands that classes of mankind shall be relieved from restrictions imposed by the artifice of man, and not by the necessities of Nature. One of the most important, if not the most important, of all these, is that which daily threatens to become the "irrepressible" woman question. What social and political rights have women? What ought they to be allowed, or not allowed,

to do, be, and suffer? And, as involved in, and underlying all these questions, how ought they to be educated?

There are philogynists as fanatical as any "misogu-
nists" who, reversing our antiquated notions, bid the
man look upon the woman as the higher type of
humanity; who ask us to regard the female intellect as
the clearer and the quicker, if not the stronger; who
desire us to look up to the feminine moral sense as the
purer and the nobler; and bid man abdicate his usurped
sovereignty over Nature in favour of the female line.
On the other hand, there are persons not to be outdone
in all loyalty and just respect for woman-kind, but by
nature hard of head and haters of delusion, however
charming, who not only repudiate the new woman-
worship which so many sentimentalists and some philosophers are desirous of setting up, but, carrying their
audacity further, deny even the natural equality of the
sexes. They assert, on the contrary, that in every
excellent character, whether mental or physical, the
average woman is inferior to the average man, in the
sense of having that character less in quantity, and lower
in quality. Tell these persons of the rapid perceptions and
the instinctive intellectual insight of women, and they
reply that the feminine mental peculiarities, which pass
under these names, are merely the outcome of a greater
impressibility to the superficial aspects of things, and of
the absence of that restraint upon expression, which, in
men, is imposed by reflection and a sense of responsibility.
Talk of the passive endurance of the weaker sex, and
opponents of this kind remind you that Job was a man,
and that, until quite recent times, patience and long-
suffering were not counted among the specially feminine

virtues. Claim passionate tenderness as especially feminine, and the inquiry is made whether all the best love-poetry in existence (except, perhaps, the "Sonnets from the Portuguese") has not been written by men; whether the song which embodies the ideal of pure and tender passion—Adelaida—was written by *Frau* Beethoven; whether it was the Fornarina, or Raphael, who painted the Sistine Madonna. Nay, we have known one such heretic go so far as to lay his hands upon the ark itself, so to speak, and to defend the startling paradox that, even in physical beauty, man is the superior. He admitted, indeed, that there was a brief period of early youth when it might be hard to say whether the prize should be awarded to the graceful undulations of the female figure, or the perfect balance and supple vigour of the male frame. But while our new Paris might hesitate between the youthful Bacchus and the Venus emerging from the foam, he averred that, when Venus and Bacchus had reached thirty, the point no longer admitted of a doubt; the male form having then attained its greatest nobility, while the female is far gone in decadence; and that, at this epoch, womanly beauty, so far as it is independent of grace or expression, is a question of drapery and accessories.

Supposing, however, that all these arguments have a certain foundation; admitting for a moment, that they are comparable to those by which the inferiority of the negro to the white man may be demonstrated, are they of any value as against woman-emancipation? Do they afford us the smallest ground for refusing to educate women as well as men—to give women the same civil and political rights as men? No mistake is so commonly

made by clever people as that of assuming a cause to be
bad because the arguments of its supporters are, to a
great extent, nonsensical.   And we conceive that those
who may laugh at the arguments of the extreme
philogynists, may yet feel bound to work heart and soul
towards the attainment of their practical ends.

As regards education, for example.   Granting the
alleged defects of women, is it not somewhat absurd to
sanction and maintain a system of education which
would seem to have been specially contrived to exaggerate all these defects ?

Naturally not so firmly strung, nor so well balanced,
as boys, girls are in great measure debarred from the
sports and physical exercises which are justly thought
absolutely necessary for the full development of the
vigour of the more favoured sex.   Women are, by nature,
more excitable than men—prone to be swept by tides of
emotion, proceeding from hidden and inward, as well as
from obvious and external causes ; and female education
does its best to weaken every physical counterpoise to
this nervous mobility—tends in all ways to stimulate the
emotional part of the mind and stunt the rest.   We find
girls naturally timid, inclined to dependence, born con-
servatives ;   and we teach them that independence is
unladylike ; that blind faith is the right frame of mind ;
and that whatever we may be permitted, and indeed
encouraged, to do to our brother, our sister is to be left
to the tyranny of authority and tradition.   With few
insignificant exceptions, girls have been educated either
to be drudges, or toys, beneath man ; or a sort of angels
above him; the highest ideal aimed at oscillating between
Clärchen and Beatrice.   The possibility that the ideal of

womanhood lies neither in the fair saint, nor in the fair
sinner; that the female type of character is neither
better nor worse than the male, but only weaker; that
women are meant neither to be men's guides nor their
playthings, but their comrades, their fellows and their
equals, so far as Nature puts no bar to that equality, does
not seem to have entered into the minds of those who
have had the conduct of the education of girls.

If the present system of female education stands self-
condemned, as inherently absurd; and if that which we
have just indicated is the true position of woman, what
is the first step towards a better state of things? We
reply, emancipate girls. Recognise the fact that they
share the senses, perceptions, feelings, reasoning powers,
emotions, of boys, and that the mind of the average girl
is less different from that of the average boy, than the
mind of one boy is from that of another; so that whatever argument justifies a given education for all boys,
justifies its application to girls as well. So far from
imposing artificial restrictions upon the acquirement of
knowledge by women, throw every facility in their way.
Let our Faustinas, if they will, toil through the whole
round of

> "Juristerei und Medizin,
> Und leider! auch Philosophie."

Let us have "sweet girl graduates" by all means. They
will be none the less sweet for a little wisdom; and the
"golden hair" will not curl less gracefully outside the
head by reason of there being brains within. Nay, if
obvious practical difficulties can be overcome, let those
women who feel inclined to do so descend into the
gladiatorial arena of life, not merely in the guise of

retiariæ, as heretofore, but as bold *sicariæ,* breasting the open fray. Let them, if they so please, become merchants, barristers, politicians. Let them have a fair field, but let them understand, as the necessary correlative, that they are to have no favour. Let Nature alone sit high above the lists, " rain influence and judge the prize."

And the result? For our parts, though loth to prophesy, we believe it will be that of other emancipations. Women will find their place, and it will neither be that in which they have been held, nor that to which some of them aspire. Nature's old salique law will not be repealed, and no change of dynasty will be effected. The big chests, the massive brains, the vigorous muscles and stout frames, of the best men will carry the day, whenever it is worth their while to contest the prizes of life with the best women. And the hardship of it is, that the very improvement of the women will lessen their chances. Better mothers will bring forth better sons, and the impetus gained by the one sex will be transmitted, in the next generation, to the other. The most Darwinian of theorists will not venture to propound the doctrine, that the physical disabilities under which women have hitherto laboured, in the struggle for existence with men, are likely to be removed by even the most skilfully conducted process of educational selection.

We are, indeed, fully prepared to believe that the bearing of children may, and ought, to become as free from danger and long disability, to the civilized woman, as it is to the savage ; nor is it improbable that, as society advances towards its right organization, motherhood will occupy a less space of woman's life than it has hitherto done.   But still, unless the human species is to come to an end altogether—a consummation which can hardly be desired by even the most ardent advocate of "women's rights"—somebody must be good enough to take the trouble and responsibility of annually adding to the world exactly as many people as die out of it.- In consequence of some domestic difficulties, Sydney Smith is said to have suggested that it would have been good for the human race had the model offered by the hive been followed, and had all the working part of the female community been neuters.   Failing any thorough-going reform of this kind, we see nothing for it but the old division of humanity into men potentially, or actually, fathers, and women potentially, if not actually, mothers. And we fear that so long as this potential motherhood is her lot, woman will be found to be fearfully weighted in the race of life.

The duty of man is to see that not a grain is piled upon that load beyond what Nature imposes; that injustice is not added to inequality.

III.

A LIBERAL EDUCATION; AND WHERE TO FIND IT.

THE business which the South London Working Men's College has undertaken is a great work; indeed, I might say, that Education, with which that college proposes to grapple, is the greatest work of all those which lie ready to a man's hand just at present.

And, at length, this fact is becoming generally recognised. You cannot go anywhere without hearing a buzz of more or less confused and contradictory talk on this subject—nor can you fail to notice that, in one point at any rate, there is a very decided advance upon like discussions in former days. Nobody outside the agricultural interest now dares to say that education is a bad thing. If any representative of the once large and powerful party, which, in former days, proclaimed this opinion, still exists in a semi-fossil state, he keeps his thoughts to himself. In fact, there is a chorus of voices, almost distressing in their harmony, raised in favour of the doctrine that education is the great panacea for human troubles, and that, if the country is not shortly to go to the dogs, everybody must be educated.

The politicians tell us, " you must educate the masses
because they are going to be masters." The clergy join
in the cry for education, for they affirm that the people
are drifting away from church and chapel into the
broadest infidelity. The manufacturers and the capitalists swell the chorus lustily. They declare that ignorance makes bad workmen; that England will soon be
unable to turn out cotton goods, or steam engines,
cheaper than other people; and then, Ichabod! Ichabod!
the glory will be departed from us. And a few voices
are lifted up in favour of the doctrine that the masses
should be educated because they are men and women
with unlimited capacities of being, doing, and suffering,
and that it is as true now, as ever it was, that the people
perish for lack of knowledge.

These members of the minority, with whom I confess
I have a good deal of sympathy, are doubtful whether
any of the other reasons urged in favour of the education
of the people are of much value—whether, indeed, some
of them are based upon either wise or noble grounds of
action. They question if it be wise to tell people that
you will do for them, out of fear of their power, what
you have left undone, so long as your only motive was
compassion for their weakness and their sorrows. And, if
ignorance of everything which it is needful a ruler should
know is likely to do so much harm in the governing
classes of the future, why is it, they ask reasonably
enough, that such ignorance in the governing classes of
the past has not been viewed with equal horror?

Compare the average artisan and the average country
squire, and it may be doubted if you will find a pin to
choose between the two in point of ignorance, class

feeling, or prejudice. It is true that the ignorance is of a different sort—that the class feeling is in favour of a different class, and that the prejudice has a distinct flavour of wrong-headedness in each case—but it is questionable if the one is either a bit better, or a bit worse, than the other. The old protectionist theory is the doctrine of trades unions as applied by the squires, and the modern trades unionism is the doctrine of the squires applied by the artisans. Why should we be worse off under one *régime* than under the other?

Again, this sceptical minority asks the clergy to think whether it is really want of education which keeps the masses away from their ministrations—whether the most completely educated men are not as open to reproach on this score as the workmen; and whether, perchance, this may not indicate that it is not education which lies at the bottom of the matter?

Once more, these people, whom there is no pleasing, venture to doubt whether the glory, which rests upon being able to undersell all the rest of the world, is a very safe kind of glory—whether we may not purchase it too dear; especially if we allow education, which ought to be directed to the making of men, to be diverted into a process of manufacturing human tools, wonderfully adroit in the exercise of some technical industry, but good for nothing else.

And, finally, these people inquire whether it is the masses alone who need a reformed and improved education. They ask whether the richest of our public schools might not well be made to supply knowledge, as well as gentlemanly habits, a strong class feeling, and eminent proficiency in cricket. They seem to think that the noble foundations of our old universities are hardly fulfilling their functions in their present posture of half-clerical seminaries, half racecourses, where men are trained to win a senior wranglership, or a double-first, as horses are trained to win a cup, with as little reference to the needs of after-life in the case of the man as in that of the racer. And, while as zealous for education as the rest, they affirm that, if the education of the richer classes were such as to fit them to be the leaders and the governors of the poorer; and, if the education of the poorer classes were such as to enable them to appreciate really wise guidance and good governance; the politicians need not fear mob-law, nor the clergy lament their want of flocks, nor the capitalists prognosticate the annihilation of the prosperity of the country.

Such is the diversity of opinion upon the why and the wherefore of education. And my hearers will be prepared to expect that the practical recommendations which are put forward are not less discordant. There is a loud cry for compulsory education. We English, in spite of constant experience to the contrary, preserve a touching faith in the efficacy of acts of parliament; and I believe we should have compulsory education in the course of next session, if there were the least probability that half a dozen leading statesmen of different parties would agree what that education should be.

Some hold that education without theology is worse than none. Others maintain, quite as strongly, that education with theology is in the same predicament. But this is certain, that those who hold the first opinion can by no means agree what theology should be taught; and that those who maintain the second are in a small minority.

At any rate "make people learn to read, write, and cipher," say a great many; and the advice is undoubtedly sensible as far as it goes. But, as has happened to me in former days, those who, in despair of getting anything better, advocate this measure, are met with the objection that it is very like making a child practise the use of a knife, fork, and spoon, without giving it a particle of meat. I really don't know what reply is to be made to such an objection.

But it would be unprofitable to spend more time in disentangling, or rather in showing up the knots in, the ravelled skeins of our neighbours. Much more to the purpose is it to ask if we possess any clue of our own which may guide us among these entanglements. And by way of a beginning, let us ask ourselves—What is education? Above all things, what is our ideal of a thoroughly liberal education?—of that education which, if we could begin life again, we would give ourselves— of that education which, if we could mould the fates to our own will, we would give our children. Well, I know not what may be your conceptions upon this matter, but I will tell you mine, and I hope I shall find that our views are not very discrepant.

Suppose it were perfectly certain that the life and fortune of every one of us would, one day or other, depend upon his winning or losing a game at chess. Don't you think that we should all consider it to be a primary duty to learn at least the names and the moves of the pieces; to have a notion of a gambit, and a keen eye for all the means of giving and getting out of check? Do you not think that we should look with a disapprobation amounting to scorn, upon the father who allowed his son, or the state which allowed its members, to grow up without knowing a pawn from a knight?

Yet it is a very plain and elementary truth, that the life, the fortune, and the happiness of every one of us, and, more or less, of those who are connected with us, do depend upon our knowing something of the rules of a game infinitely more difficult and complicated than chess. It is a game which has been played for untold ages, every man and woman of us being one of the two players in a game of his or her own. The chess-board is the world, the pieces are the phenomena of the universe, the rules of the game are what we call the laws of Nature. The player on the other side is hidden from us. We know that his play is always fair, just, and patient. But also we know, to our cost, that he never overlooks a mistake, or makes the smallest allowance for ignorance. To the man who plays well, the highest stakes are paid, with that sort of overflowing generosity with which the strong shows delight in strength. And one who plays ill is checkmated—without haste, but without remorse.

My metaphor will remind some of you of the famous picture in which Retzsch has depicted Satan playing at chess with man for his soul. Substitute for the mocking fiend in that picture, a calm, strong angel who is playing for love, as we say, and would rather lose than win—and I should accept it as an image of human life.

Well, what I mean by Education is learning the rules of this mighty game. In other words, education is the instruction of the intellect in the laws of Nature, under which name I include not merely things and their forces, but men and their ways; and the fashioning of the

affections and of the will into an earnest and loving
desire to move in harmony with those laws. For me,
education means neither more nor less than this. Anything which professes to call itself education must be
tried by this standard, and if it fails to stand the test, I
will not call it education, whatever may be the force of
authority, or of numbers, upon the other side.

It is important to remember that, in strictness, there
is no such thing as an uneducated man. Take an extreme case. Suppose that an adult man, in the full
vigour of his faculties, could be suddenly placed in the
world, as Adam is said to have been, and then left to
do as he best might. How long would he be left
uneducated? Not five minutes. Nature would begin
to teach him, through the eye, the ear, the touch, the
properties of objects. Pain and pleasure would be at his
elbow telling him to do this and avoid that; and by slow
degrees the man would receive an education, which, if
narrow, would be thorough, real, and adequate to his
circumstances, though there would be no extras and very
few accomplishments.

And if to this solitary man entered a second Adam,
or, better still, an Eve, a new and greater world, that of
social and moral phenomena, would be revealed. Joys
and woes, compared with which all others might seem
but faint shadows, would spring from the new relations.
Happiness and sorrow would take the place of the
coarser monitors, pleasure and pain; but conduct would
still be shaped by the observation of the natural consequences of actions; or, in other words, by the laws of
the nature of man.

To every one of us the world was once as fresh and

new as to Adam. And then, long before we were susceptible of any other mode of instruction, Nature took
us in hand, and every minute of waking life brought its
educational influence, shaping our actions into rough
accordance with Nature's laws, so that we might not be
ended untimely by too gross disobedience. Nor should
I speak of this process of education as past, for any one,
be he as old as he may. For every man, the world is as
fresh as it was at the first day, and as full of untold
novelties for him who has the eyes to see them. And
Nature is still continuing her patient education of us in
that great university, the universe, of which we are all
members—Nature having no Test-Acts.

Those who take honours in Nature's university, who
learn the laws which govern men and things and obey
them, are the really great and successful men in this
world. The great mass of mankind are the " Poll," who
pick up just enough to get through without much dis-
credit. Those who won't learn at all are plucked; and
then you can't come up again. Nature's pluck means
extermination.

Thus the question of compulsory education is settled
so far as Nature is concerned. Her bill on that question
was framed and passed long ago. But, like all compulsory legislation, that of Nature is harsh and wasteful
in its operation. Ignorance is visited as sharply as
wilful disobedience—incapacity meets with the same
punishment as crime. Nature's discipline is not even a
word and a blow, and the blow first; but the blow
without the word. It is left to you to find out why
your ears are boxed.

The object of what we commonly call education—that

education in which man intervenes and which I shall distinguish as artificial education—is to make good these defects in Nature's methods; to prepare the child to receive Nature's education, neither incapably nor ignorantly, nor with wilful disobedience; and to understand the preliminary symptoms of her displeasure, without waiting for the box on the ear. In short, all artificial education ought to be an anticipation of natural education. And a liberal education is an artificial education, which has not only prepared a man to escape the great evils of disobedience to natural laws, but has trained him to appreciate and to seize upon the rewards, which Nature scatters with as free a hand as her penalties.

That man, I think, has had a liberal education, who has been so trained in youth that his body is the ready servant of his will, and does with ease and pleasure all the work that, as a mechanism, it is capable of; whose intellect is a clear, cold, logic engine, with all its parts of equal strength, and in smooth working order; ready, like a steam engine, to be turned to any kind of work, and spin the gossamers as well as forge the anchors of the mind; whose mind is stored with a knowledge of the great and fundamental truths of Nature and of the laws of her operations; one who, no stunted ascetic, is full of life and fire, but whose passions are trained to come to heel by a vigorous will, the servant of a tender conscience; who has learned to love all beauty, whether of Nature or of art, to hate all vileness, and to respect others as himself.

Such an one and no other, I conceive, has had a liberal education; for he is, as completely as a man can be, in harmony with Nature. He will make the best of her, and she of him. They will get on together rarely ; she as his ever beneficent mother ; he as her mouth-piece, her conscious self, her minister and interpreter.

Where is such an education as this to be had ? Where is there any approximation to it ? Has any one tried to found such an education ? Looking over the length and breadth of these islands, I am afraid that all these questions must receive a negative answer. Consider our primary schools, and what is taught in them. A child learns :—

1. To read, write, and cipher, more or less well; but in a very large proportion of cases not so well as to take pleasure in reading, or to be able to write the commonest letter properly.

2. A quantity of dogmatic theology, of which the child, nine times out of ten, understands next to nothing.

3. Mixed up with this, so as to seem to stand or fall with it, a few of the broadest and simplest principles of morality. This, to my mind, is much as if a man of science should make the story of the fall of the apple in Newton's garden, an integral part of the doctrine of gravitation, and teach it as of equal authority with the law of the inverse squares.

4. A good deal of Jewish history and Syrian geography, and, perhaps, a little something about English history and the geography of the child's own country. But I doubt if there is a primary school in England in which hangs a map of the hundred in which the village lies, so that the children may be practically taught by it what a map means.

5. A certain amount of regularity, attentive obedience, respect for others : obtained by fear, if the master be incompetent or foolish; by love and reverence, if he be wise.

So far as this school course embraces a training in the theory and practice of obedience to the moral laws of Nature, I gladly admit, not only that it contains a valuable educational element, but that, so far, it deals with the most valuable and important part of all education. Yet, contrast what is done in this direction with what might be done; with the time given to matters of comparatively no importance; with the absence of any attention to things of the highest moment; and one is tempted to think of Falstaff's bill and "the halfpenny worth of bread to all that quantity of sack."

Let us consider what a child thus "educated" knows, and what it does not know. Begin with the most important topic of all—morality, as the guide of conduct. The child knows well enough that some acts meet with approbation and some with disapprobation. But it has never heard that there lies in the nature of things a reason for every moral law, as cogent and as well defined as that which underlies every physical law ; that stealing and lying are just as certain to be followed by evil consequences, as putting your hand in the fire, or jumping out of a garret window. Again, though the scholar may have been made acquainted, in dogmatic fashion, with the broad laws of morality, he has had no training in the application of those laws to the difficult problems which result from the complex conditions of modern civilization. Would it not be very hard to expect anyone to solve a problem in conic sections who had merely been taught the axioms and definitions of mathematical science?

A workman has to bear hard labour, and perhaps privation, while he sees others rolling in wealth, and feeding their dogs with what would keep his children from starvation. Would it not be well to have helped that man to calm the natural promptings of discontent by showing him, in his youth, the necessary connexion of the moral law which prohibits stealing with the stability of society—by proving to him, once for all, that it is better for his own people, better for himself, better for future generations, that he should starve than steal? If you have no foundation of knowledge, or habit of thought, to work upon, what chance have you of persuading a hungry man that a capitalist is not a thief "with a circumbendibus?" And if he honestly believes that, of what avail is it to quote the commandment against stealing, when he proposes to make the capitalist disgorge?

Again, the child learns absolutely nothing of the history or the political organization of his own country. His general impression is, that everything of much importance happened a very long while ago; and that the Queen and the gentlefolks govern the country much after the fashion of King David and the elders and nobles of Israel—his sole models. Will you give a man with this much information a vote? In easy times he sells it for a pot of beer. Why should he not? It is of about as much use to him as a chignon, and he knows as much what to do with it, for any other purpose. In bad times, on the contrary, he applies his simple theory of government, and believes that his rulers are the cause of his sufferings—a belief which sometimes bears remarkable practical fruits.

Least of all, does the child gather from this primary

"education" of ours a conception of the laws of the physical world, or of the relations of cause and effect therein. And this is the more to be lamented, as the poor are especially exposed to physical evils, and are more interested in removing them than any other class of the community. If any one is concerned in knowing the ordinary laws of mechanics one would think it is the hand-labourer, whose daily toil lies among levers and pulleys ; or among the other implements of artisan work. And if any one is interested in the laws of health, it is the poor workman, whose strength is wasted by ill-prepared food, whose health is sapped by bad ventilation and bad drainage, and half whose children are massacred by disorders which might be prevented. Not only does our present primary education carefully abstain from hinting to the workman that some of his greatest evils are traceable to mere physical agencies, which could be removed by energy, patience, and frugality ; but it does worse— it renders him, so far as it can, deaf to those who could help him, and tries to substitute an Oriental submission to what is falsely declared to be the will of God, for his natural tendency to strive after a better condition.

What wonder then, if very recently, an appeal has been made to statistics for the profoundly foolish purpose of showing that education is of no good—that it diminishes neither misery, nor crime, among the masses of mankind ? I reply, why should the thing which has been called education do either the one or the other ? If I am a knave or a fool, teaching me to read and write won't make me less of either one or the other—unless somebody shows me how to put my reading and writing to wise and good purposes.

Suppose any one were to argue that medicine is of no use, because it could be proved statistically, that the percentage of deaths was just the same, among people who had been taught how to open a medicine chest, and among those who did not so much as know the key by sight. The argument is absurd; but it is not more preposterous than that against which I am contending. The only medicine for suffering, crime, and all the other woes of mankind, is wisdom. Teach a man to read and write, and you have put into his hands the great keys of the wisdom box. But it is quite another matter whether he ever opens the box or not. And he is as likely to poison as to cure himself, if, without guidance, he swallows the first drug that comes to hand. In these times a man may as well be purblind, as unable to read —lame, as unable to write. But I protest that, if I thought the alternative were a necessary one, I would rather that the children of the poor should grow up ignorant of both these mighty arts, than that they should remain ignorant of that knowledge to which these arts are means.

It may be said that all these animadversions may apply to primary schools, but that the higher schools, at any rate, must be allowed to give a liberal education. In fact, they professedly sacrifice everything else to this object.

Let us inquire into this matter. What do the higher schools, those to which the great middle class of the country sends it children, teach, over and above the instruction given in the primary schools? There is a little more reading and writing of English. But, for all that, every one knows that it is a rare thing to find a boy of the middle or upper classes who can read aloud decently, or who can put his thoughts on paper in clear and grammatical (to say nothing of good or elegant) language. The "ciphering" of the lower schools expands into elementary mathematics in the higher; into arithmetic, with a little algebra, a little Euclid. But I doubt if one boy in five hundred has ever heard the explanation of a rule of arithmetic, or knows his Euclid otherwise than by rote.

Of theology, the middle class schoolboy gets rather less than poorer children, less absolutely and less relatively, because there are so many other claims upon his attention. I venture to say that, in the great majority of cases, his ideas on this subject when he leaves school are of the most shadowy and vague description, and associated with painful impressions of the weary hours spent in learning collects and catechism by heart.

Modern geography, modern history, modern literature; the English language as a language; the whole circle of the sciences, physical, moral, and social, are even more completely ignored in the higher than in the lower schools. Up till within a few years back, a boy might have passed through any one of the great public schools with the greatest distinction and credit, and might never so much as have heard of one of the subjects I have just mentioned. He might never have heard that the earth goes round the sun; that England underwent a great revolution in 1688, and France another in 1789; that there once lived certain notable men called Chaucer, Shakspeare, Milton, Voltaire, Goethe, Schiller. The first might be a German and the last an Englishman for anything he could tell you to the contrary. And as for science, the only idea the word would suggest to his mind would be dexterity in boxing.

I have said that this was the state of things a few years back, for the sake of the few righteous who are to be found among the educational cities of the plain. But I would not have you too sanguine about the result, if you sound the minds of the existing generation of public school-boys, on such topics as those I have mentioned.

Now let us pause to consider this wonderful state of affairs ; for the time will come when Englishmen will quote it as the stock example of the stolid stupidity of their ancestors in the nineteenth century. The most thoroughly commercial people, the greatest voluntary wanderers and colonists the world has ever seen, are precisely the middle classes of this country. If there be a people which has been busy making history on the great scale for the last three hundred years—and the most profoundly interesting history—history which, if it happened to be that of Greece or Rome, we should study with avidity—it is the English. If there be a people which, during the same period, has developed a remarkable literature, it is our own. If there be a nation whose prosperity depends absolutely and wholly upon their mastery over the forces of Nature, upon their intelligent apprehension of, and obedience to, the laws of the creation and distribution of wealth, and of the stable equilibrium of the forces of society, it is pre- cisely this nation. And yet this is what these wonderful people tell their sons :—" At the cost of from one to two thousand pounds of our hard earned money, we devote twelve of the most precious years of your lives to school. There you shall toil, or be supposed to toil; but there you shall not learn one single thing of all those you will most want to know, directly you leave school and enter upon the practical business of life. You will in all probability go into business, but you shall not know where, or how, any article of commerce is produced, or the difference between an export or an import, or the meaning of the word 'capital.' You will very likely settle in a colony, but you shall not know whether Tasmania is part of New South Wales, or *vice versâ*.

" Very probably you may become a manufacturer, but you shall not be provided with the means of understanding the working of one of your own steam-engines, or the nature of the raw products you employ; and, when you are asked to buy a patent, you shall not have the slightest means of judging whether the inventor is an impostor who is contravening the elementary principles of science, or a man who will make you as rich as Crœsus.

" You will very likely get into the House of Commons. You will have to take your share in making laws which may prove a blessing or a curse to millions of men. But you shall not hear one word respecting the political organization of your country; the meaning of the controversy between freetraders and protectionists shall never have been mentioned to you; you shall not so much as know that there are such things as economical laws.

" The mental power which will be of most importance in your daily life will be the power of seeing things as they are without regard to authority; and of drawing accurate general conclusions from particular facts. But at school and at college you shall know of no source of truth but authority ; nor exercise your reasoning faculty upon anything but deduction from that which is laid down by authority.

" You will have to weary your soul with work, and many a time eat your bread in sorrow and in bitterness, and you shall not have learned to take refuge in the great source of pleasure without alloy, the serene resting-place for worn human nature,—the world of art."

Said I not rightly that we are a wonderful people ? I am quite prepared to allow, that education entirely devoted to these omitted subjects might not be a completely liberal education. But is an education which ignores them all, a liberal education ? Nay, is it too much to say that the education which should embrace these subjects and no others, would be a real education, though an incomplete one ; while an education which omits them is really not an education at all, but a more or less useful course of intellectual gymnastics ?

For what does the middle-class school put in the place of all these things which are left out ? It substitutes what is usually comprised under the compendious title of the " classics "—that is to say, the languages, the literature, and the history of the ancient Greeks and Romans, and the geography of so much of the world as was known to these two great nations of antiquity. Now, do not expect me to depreciate the earnest and enlightened pursuit of classical learning. I have not the least desire to speak ill of such occupations, nor any sympathy with those who run them down. - On

the contrary, if my opportunities had lain in that direction, there is no investigation into which I could
have thrown myself with greater delight than that of
antiquity.

What science can present greater attractions than
philology? How can a lover of literary excellence fail
to rejoice in the ancient masterpieces? And with what
consistency could I, whose business lies so much in the
attempt to decipher the past, and to build up intelligible
forms out of the scattered fragments of long-extinct
beings, fail to take a sympathetic, though an unlearned,
interest in the labours of a Niebuhr, a Gibbon, or a
Grote? Classical history is a great section of the palæontology of man; and I have the same double respect
for it as for other kinds of palæontology—that is to say,
a respect for the facts which it establishes as for all
facts, and a still greater respect for it as a preparation
for the discovery of a law of progress.

But if the classics were taught as they might be
taught—if boys and girls were instructed in Greek and
Latin, not merely as languages, but as illustrations of
philological science; if a vivid picture of life on the
shores of the Mediterranean, two thousand years ago,
were imprinted on the minds of scholars; if ancient
history were taught, not as a weary series of feuds and
fights, but traced to its causes in such men placed under
such conditions; if, lastly, the study of the classical
books were followed in such a manner as to impress boys
with their beauties, and with the grand simplicity of
their statement of the everlasting problems of human
life, instead of with their verbal and grammatical peculiarities; I still think it as little proper that they should

E

form the basis of a liberal education for our contemporaries, as I should think it fitting to make that sort of palæontology with which I am familiar, the back-bone of modern education.

It is wonderful how close a parallel to classical training could be made out of that palæontology to which I refer. In the first place I could get up an osteological primer so arid, so pedantic in its terminology, so altogether distasteful to the youthful mind, as to beat the recent famous production of the head-masters out of the field in all these excellences. Next, I could exercise my boys upon easy fossils, and bring out all their powers of memory and all their ingenuity in the application of my osteo-grammatical rules to the interpretation, or construing, of those fragments. To those who had reached the higher classes, I might supply odd bones to be built up into animals, giving great honour and reward to him who succeeded in fabricating monsters most entirely in accordance with the rules. That would answer to verse-making and essay-writing in the dead languages.

To be sure, if a great comparative anatomist were to look at these fabrications he might shake his head, or laugh. But what then? Would such a catastrophe destroy the parallel? What think you would Cicero, or Horace, say to the production of the best sixth form going? And would not Terence stop his ears and run out if he could be present at an English performance of his own plays? Would Hamlet, in the mouths of a set of French actors, who should insist on pronouncing English after the fashion of their own tongue, be more hideously ridiculous?

But it will be said that I am forgetting the beauty, and the human interest, which appertain to classical studies. To this I reply that it is only a very strong man who can appreciate the charms of a landscape, as he is toiling up a steep hill, along a bad road. What with short-windedness, stones, ruts, and a pervading sense of the wisdom of rest and be thankful, most of us have little enough sense of the beautiful under these circumstances. The ordinary school-boy is precisely in this case. He finds Parnassus uncommonly steep, and there is no chance of his having much time or inclination to look about him till he gets to the top. And nine times out of ten he does not get to the top.

But if this be a fair picture of the results of classical teaching at its best—and I gather from those who have authority to speak on such matters that it is so— what is to be said of classical teaching at its worst, or in other words, of the classics of our ordinary middle-class schools?[1] I will tell you. It means getting up endless forms and rules by heart. It means turning Latin and Greek into English, for the mere sake of being able to do it, and without the smallest regard to the worth, or worthlessness, of the author read. It means the learning of innumerable, not always decent, fables in such a shape that the meaning they once had is dried up into utter trash; and the only impression left upon a boy's mind is, that the people who believed such things must have been the greatest idiots the world ever saw. And it means, finally, that after a dozen years spent at this kind of work, the sufferer

---

[1] For a justification of what is here said about these schools, see that valuable book, "Essays on a Liberal Education," *passim*.

shall be incompetent to interpret a passage in an author he has not already got up ; that he shall loathe the sight of a Greek or Latin book ; and that he shall never open, or think of, a classical writer again, until, wonderful to relate, he insists upon submitting his sons to the same process.

These be your gods, O Israel ! For the sake of this net result (and respectability) the British father denies his children all the knowledge they might turn to account in life, not merely for the achievement of vulgar success, but for guidance in the great crises of human existence. This is the stone he offers to those whom he is bound by the strongest and tenderest ties to feed with bread.

If primary and secondary education are in this unsatisfactory state, what is to be said to the universities ? This is an awful subject, and one I almost fear to touch with my unhallowed hands ; but I can tell you what those say who have authority to speak.

The Rector of Lincoln College, in his lately published, valuable " Suggestions for Academical Organization with especial reference to Oxford," tells us (p. 127) :—

"The colleges were, in their origin, endowments, not for the elements of a general liberal education, but for the prolonged study of special and professional faculties by men of riper age. The universities embraced both these objects. The colleges, while they incidentally aided in elementary education, were specially devoted to the highest learning. . . . .

"This was the theory of the middle-age university and the design of collegiate foundations in their origin. Time and circumstances have brought about a total change. The colleges no longer promote the researches of science, or direct professional study. Here and there college walls may shelter an occasional student, but not in larger proportions than may be found in private life. Elementary teaching of youths under twenty is now the only function performed by the university, and almost the only object of college endowments. Colleges were homes for the life-study of the highest and most abstruse parts of knowledge. They have become boarding schools in which the elements of the learned languages are taught to youths."

If Mr. Pattison's high position, and his obvious love and respect for his university, be insufficient to convince the outside world that language so severe is yet no more than just, the authority of the Commissioners who reported on the University of Oxford in 1850 is open to no challenge. Yet they write :—

"It is generally acknowledged that both Oxford and the country at large suffer greatly from the absence of a body of learned men devoting their lives to the cultivation of science, and to the direction of academical education.

"The fact that so few books of profound research emanate from the University of Oxford, materially impairs its character as a seat of learning, and consequently its hold on the respect of the nation."

Cambridge can claim no exemption from the reproaches addressed to Oxford. And thus there seems no escape from the admission that what we fondly call our great seats of learning are simply "boarding schools" for bigger boys ; that learned men are not more numerous in them than out of them ; that the advancement of knowledge is not the object of fellows of colleges ; that, in the philosophic calm and meditative stillness of their greenswarded courts, philosophy does not thrive, and meditation bears few fruits.

It is my great good fortune to reckon amongst my friends resident members of both universities, who are men of learning and research, zealous cultivators of science, keeping before their minds a noble ideal of a university, and doing their best to make that ideal a reality ; and, to me, they would necessarily typify the universities, did not the authoritative statements I have quoted compel me to believe that they are exceptional, and not representative men. Indeed, upon calm consideration, several circumstances lead me to think that the Rector of Lincoln College and the Commissioners cannot be far wrong.

I believe there can be no doubt that the foreigner who should wish to become acquainted with the scientific, or the literary, activity of modern England, would simply lose his time and his pains if he visited our universities with that object.

And, as for works of profound research on any subject, and, above all, in that classical lore for which the universities profess to sacrifice almost everything else, why, a third-rate, poverty-stricken German university turns out more produce of that kind in one year, than our vast and wealthy foundations elaborate in ten.

Ask the man who is investigating any question, profoundly and thoroughly—be it historical, philosophical, philological, physical, literary, or theological ; who is trying to make himself master of any abstract subject (except, perhaps, political economy and geology, both

of which are intensely Anglican sciences) whether he
is not compelled to read half a dozen times as many
German, as English, books ?   And whether, of these
English books, more than one in ten is the work of
a fellow of a college, or a professor of an English
university ?

Is this from any lack of power in the English as
compared with the German mind ? The countrymen
of Grote and of Mill, of Faraday, of Robert Brown,
of Lyell, and of Darwin, to go no further back than
the contemporaries of men of middle age, can afford
to smile at such a suggestion.   England can show now,
as she has been able to show in every generation since
civilization spread over the West, individual men who
hold their own against the world, and keep alive the
old tradition of her intellectual eminence.

But, in the majority of cases, these men are what
they are in virtue of their native intellectual force, and
of a strength of character which will not recognise impedi-
ments.   They are not trained in the courts of the
Temple of Science, but storm the walls of that edifice in
all sorts of irregular ways, and with much loss of time
and power, in order to obtain their legitimate positions.

Our universities not only do not encourage such men ;
do not offer them positions, in which it should be their
highest duty to do, thoroughly, that which they are most
capable of doing ; but, as far as possible, university training shuts out of the minds of those among them, who
are subjected to it, the prospect that there is anything in
the world for which they are specially fitted.   Imagine
the success of the attempt to still the intellectual hunger
of any of the men I have mentioned, by putting before

him, as the object of existence, the successful mimicry of the measure of a Greek song, or the roll of Ciceronian prose! Imagine how much success would be likely to attend the attempt to persuade such men, that the education which leads to perfection in such elegancies is alone to be called culture; while the facts of history, the process of thought, the conditions of moral and social existence, and the laws of physical nature, are left to be dealt with as they may, by outside barbarians!

It is not thus that the German universities, from being beneath notice a century ago, have become what they are now—the most intensely cultivated and the most productive intellectual corporations the world has ever seen.

The student who repairs to them sees in the list of classes and of professors a fair picture of the world of knowledge. Whatever he needs to know there is some one ready to teach him, some one competent to discipline him in the way of learning; whatever his special bent, let him but be able and diligent, and in due time he shall find distinction and a career. Among his professors, he sees men whose names are known and revered throughout the civilized world; and their living example infects him with a noble ambition, and a love for the spirit of work.

The Germans dominate the intellectual world by virtue of the same simple secret as that which made Napoleon the master of old Europe. They have declared *la carrière ouverte aux talents*, and every Bursch marches with a professor's gown in his knapsack. Let him become a great scholar, or man of science, and ministers will compete for his services. In Germany, they do not leave the chance of his holding the office he would render illustrious to the tender mercies of a hot canvass, and the final wisdom of a mob of country parsons.

In short, in Germany, the universities are exactly what the Rector of Lincoln and the Commissioners tell us the English universities are not ; that is to say, corporations " of learned men devoting their lives to the cultivation of science, and the direction of academical education." They are not " boarding schools for youths," nor clerical seminaries ; but institutions for the higher culture of men, in which the theological faculty is of no more importance, or prominence, than the rest ; and which are truly " universities," since they strive to represent and embody the totality of human knowledge, and to find room for all forms of intellectual activity.

May zealous and clear-headed reformers like Mr. Pattison succeed in their noble endeavours to shape our universities towards some such ideal as this, without losing what is valuable and distinctive in their social tone ! But until they have succeeded, a liberal education will be no more obtainable in our Oxford and Cambridge Universities than in our public schools.

If I am justified in my conception of the ideal of a liberal education ; and if what I have said about the existing educational institutions of the country is also true, it is clear that the two have no sort of relation to one ·another ; that the best of our schools and the most complete of our university trainings give but a narrow, one-sided, and essentially illiberal education— while the worst give what is really next to no education at all.   The South London Working-Men's College could not copy any of these institutions if it would. I am bold enough to express the conviction that it ought not if it could.

For what is wanted is the reality and not the mere name of a liberal education; and this College must steadily set before itself the ambition to be able to give that education sooner or later.   At present we are but beginning, sharpening our educational tools, as it were, and, except a modicum of physical science, we are not able to offer much more than is to be found in an ordinary school.

Moral and social science—one of the greatest and most fruitful of our future classes, I hope—at present lacks only one thing in our programme, and that is a teacher.   A considerable want, no doubt; but it must be recollected that it is much better to want a teacher than to want the desire to learn.

Further, we need what, for want of a better name, I must call Physical Geography.   What I mean is that which the Germans call "*Erdkunde.*"   It is a description of the earth, of its place and relation to other bodies; of its general structure, and of its great features —winds, tides, mountains, plains; of the chief forms of the vegetable and animal worlds, of the varieties of man.   It is the peg upon which the greatest quantity of useful and entertaining scientific information can be suspended.

Literature is not upon the College programme; but hope some day to see it there.   For literature is the greatest of all sources of refined pleasure, and one of the great uses of a liberal education is to enable us to enjoy that pleasure. There is scope enough for the purposes of liberal education in the study of the rich treasures of our own language alone. All that is needed is direction, and the cultivation of a refined taste by attention to sound criticism. But there is no reason why French and German should not be mastered sufficiently to read what is worth reading in those languages, with pleasure and with profit.

And finally, by-and-by, we must have History; treated not as a succession of battles and dynasties; not as a series of biographies; not as evidence that Providence has always been on the side of either Whigs or Tories; but as the development of man in times past, and in other conditions than our own.

But, as it is one of the principles of our College to be self-supporting, the public must lead, and we must follow, in these matters. If my hearers take to heart what I have said about liberal education, they will desire these things, and I doubt not we shall be able to supply them. But we must wait till the demand is made.

SCIENTIFIC EDUCATION: NOTES OF AN AFTER-DINNER SPEECH.

[Mr. Thackeray, talking of after-dinner speeches, has lamented that "one never can recollect the fine things one thought of in the cab," in going to the place of entertainment. I am not aware that there are any "fine things" in the following pages, but such as there are stand to a speech which really did get itself spoken, at the hospitable table of the Liverpool Philomathic Society, more or less in the position of what "one thought of in the cab."]

The introduction of scientific training into the general education of the country is a topic upon which I could not have spoken, without some more or less apologetic introduction, a few years ago. But upon this, as upon other matters, public opinion has of late undergone a rapid modification. Committees of both Houses of the Legislature have agreed that something must be done in this direction, and have even thrown out timid and faltering suggestions as to what should be done; while at the opposite pole of society, committees of working-men have expressed their conviction that scientific training is the one thing needful for their advancement, whether as men, or as workmen. Only the other day, it was my duty to take part in the reception of a deputation of London working men, who desired to learn from Sir Roderick Murchison, the Director of the Royal School of Mines, whether the organization of the Institution in Jermyn Street could be made available for the supply of that scientific instruction, the need of which could not have been apprehended, or stated, more clearly than it was by them.

The heads of colleges in our great Universities (who have not the reputation of being the most mobile of persons) have, in several cases, thought it well that, out of the great number of honours and rewards at their disposal, a few should hereafter be given to the cultivators of the physical sciences. Nay, I hear that some colleges have even gone so far as to appoint one, or, may be, two special tutors for the purpose of putting the facts and principles of physical science before the undergraduate mind. And I say it with gratitude and great respect for those eminent persons, that the head masters of our public schools, Eton, Harrow, Winchester, have addressed themselves to the problem of introducing instruction in physical science among the studies of those great educational bodies, with much honesty of purpose and enlightenment of understanding ; and I live in hope that, before long, important changes in this direction will be carried into effect in those strongholds of ancient prescription. In fact, such changes have already been made, and physical science, even now, constitutes a recognised element of the school curriculum in Harrow and Rugby, whilst

I understand that ample preparations for such studies are being made at Eton and elsewhere.

Looking at these facts, I might perhaps spare myself the trouble of giving any reasons for the introduction of physical science into elementary education; yet I cannot but think that it may be well, if I place before you some considerations which, perhaps, have hardly received full attention.

At other times, and in other places, I have endeavoured to state the higher and more abstract arguments, by which the study of physical science may be shown to be indispensable to the complete training of the human mind; but I do not wish it to be supposed that, because I happen to be devoted to more or less abstract and "unpractical" pursuits, I am insensible to the weight which ought to be attached to that which has been said to be the English conception of Paradise —"namely, getting on." I look upon it, that "getting on" is a very important matter indeed. I do not mean merely for the sake of the coarse and tangible results of success, but because humanity is so constituted that a vast number of us would never be impelled to those stretches of exertion which make us wiser and more capable men, if it were not for the absolute necessity of putting on our faculties all the strain they will bear, for the purpose of "getting on" in the most practical sense.

Now the value of a knowledge of physical science as a means of getting on, is indubitable. There are hardly any of our trades, except the merely huckstering ones, in which some knowledge of science may not be directly profitable to the pursuer of that occupation.

As industry attains higher stages of its development, as its processes become more complicated and refined, and competition more keen, the sciences are dragged in, one by one, to take their share in the fray ; and he who can best avail himself of their help is the man who will come out uppermost in that struggle for existence, which goes on as fiercely beneath the smooth surface of modern society, as among the wild inhabitants of the woods.

But, in addition to the bearing of science on ordinary practical life, let me direct your attention to its immense influence on several of the professions. I ask any one who has adopted the calling of an engineer, how much time he lost when he left school, because he had to devote himself to pursuits which were absolutely novel and strange, and of which he had not obtained the remotest conception from his instructors? He had to familiarize himself with ideas of the course and powers of Nature, to which his attention had never been directed during his school-life, and to learn, for the first time, that a world of facts lies outside and beyond the world of words. I appeal to those who know what Engineering is, to say how far I am right in respect to that profession ; but with regard to another, of no less importance, I shall venture to speak of my own knowledge. There is no one of us who may not at any moment be thrown, bound hand and foot by physical incapacity, into the hands of a medical practitioner. The chances of life and death for all and each of us may, at any moment, depend on the skill with which that practitioner is able to make out what is wrong in our bodily frames, and on his ability to apply the proper remedy to the defect.

The necessities of modern life are such, and the class from which the medical profession is chiefly recruited is so situated, that few medical men can hope to spend more than three or four, or it may be five, years in the pursuit of those studies which are immediately germane to physic. How is that all too brief period spent at present? I speak as an old examiner, having served some eleven or twelve years in that capacity in the University of London, and therefore having a practical acquaintance with the subject; but I might fortify myself by the authority of the President of the College of Surgeons, Mr. Quain, whom I heard the other day in an admirable address (the Hunterian Oration) deal fully and wisely with this very topic.[1]

---

[1] Mr. Quain's words (*Medical Times and Gazette*, February 20) are :—" A few words as to our special Medical course of instruction and the influence upon it of such changes in the elementary schools as I have mentioned. The student now enters at once upon several sciences—physics, chemistry, anatomy, physiology, botany, pharmacy, therapeutics — all these, the facts and the language and the laws of each, to be mastered in eighteen months. Up to the beginning of the Medical course many have learned little. We cannot claim anything better than the Examiner of the University of London and the Cambridge Lecturer have reported for their Universities. Supposing that at school young people had acquired some exact elementary knowledge in physics, chemistry, and a branch of natural history—say botany—with the physiology connected with it, they would then have gained necessary knowledge, with some practice in inductive reasoning. The whole studies are processes of observation and induction—the best discipline of the mind for the purposes of life—for our purposes not less than any. ' By such study (says Dr. Whewell) of one or more departments of inductive science the mind may escape from the thraldom of mere words.' By that plan the burden of the early Medical course would be much lightened, and more time devoted to practical studies, including Sir Thomas Watson's ' final and supreme stage ' of the knowledge of Medicine."

A young man commencing the study of medicine is at once required to endeavour to make an acquaintance with a number of sciences, such as Physics, as Chemistry, as Botany, as Physiology, which are absolutely and entirely strange to him, however excellent his so-called education at school may have been. Not only is he devoid of all apprehension of scientific conceptions, not only does he fail to attach any meaning to the words "matter," "force," or "law" in their scientific senses, but, worse still, he has no notion of what it is to come into contact with nature, or to lay his mind alongside of a physical fact, and try to conquer it, in the way our great naval hero told his captains to master their enemies. His whole mind has been given to books, and I am hardly exaggerating if I say that they are more real to him than Nature. He imagines that all knowledge can be got out of books, and rests upon the authority of some master or other; nor does he entertain any misgiving that the method of learning which led to proficiency in the rules of grammar, will suffice to lead him to a mastery of the laws of Nature. The youngster, thus unprepared for serious study, is turned loose among his medical studies, with the result, in nine cases out of ten, that the first year of his curriculum is spent in learning how to learn. Indeed, he is lucky, if at the end of the first year, by the exertions of his teachers and his own industry, he has acquired even that art of arts. After which there remain not more than three, or perhaps four, years for the profitable study of such vast sciences as Anatomy, Physiology, Therapeutics, Medicine, Surgery, Obstetrics, and the like, upon his knowledge or ignorance of which it depends whether

the practitioner shall diminish, or increase, the bills of
mortality. Now what is it but the preposterous condition of ordinary school education which prevents a
young man of seventeen, destined for the practice of
medicine, from being fully prepared for the study of
nature; and from coming to the medical school, equipped
with that preliminary knowledge of the principles of
Physics, of Chemistry, and of Biology, upon which he
has now to waste one of the precious years, every
moment of which ought to be given to those studies
which bear directly upon the knowledge of his
profession ?

There is another profession, to the members of which,
I think, a certain preliminary knowledge of physical
science might be quite as valuable as to the medical
man. The practitioner of medicine sets before himself
the noble object of taking care of man's bodily welfare ;
but the members of this other profession undertake to
"minister to minds diseased," and, so far as may be,
to diminish sin and soften sorrow. Like the medical
profession, the clerical, of which I now speak, rests its
power to heal upon its knowledge of the order of the
universe—upon certain theories of man's relation to
that which lies outside him. It is not my business to
express any opinion about these theories. I merely
wish to point out that, like all other theories, they are
professedly based upon matter of fact. Thus the clerical
profession has to deal with the facts of Nature from a
certain point of view ; and hence it comes into contact
with that of the man of science, who has to treat the
same facts from another point of view. You know how
often that contact is to be described as collision, or

violent friction; and how great the heat, how little the light, which commonly results from it.

In the interests of fair play, to say nothing of those of mankind, I ask, Why do not the clergy as a body acquire, as a part of their preliminary education, some such tincture of physical science as will put them in a position to understand the difficulties in the way of accepting their theories, which are forced upon the mind of every thoughtful and intelligent man, who has taken the trouble to instruct himself in the elements of natural knowledge?

Some time ago I attended a large meeting of the clergy, for the purpose of delivering an address which I had been invited to give. I spoke of some of the most elementary facts in physical science, and of the manner in which they directly contradict certain of the ordinary teachings of the clergy. The result was, that, after I had finished, one section of the assembled ecclesiastics attacked me with all the intemperance of pious zeal, for stating facts and conclusions which no competent judge doubts; while, after the first speakers had subsided, amidst the cheers of the great majority of their colleagues, the more rational minority rose to tell me that I had taken wholly superfluous pains, that they already knew all about what I had told them, and perfectly agreed with me. A hard-headed friend of mine, who was present, put the not unnatural question, "Then why don't you say so in your pulpits?" to which inquiry I heard no reply.

In fact the clergy are at present divisible into three sections: an immense body who are ignorant and speak out; a small proportion who know and are silent;

and a minute minority who know and speak according
to their knowledge. By the clergy, I mean especially
the Protestant clergy. Our great antagonist—I speak
as a man of science—the Roman Catholic Church, the
one great spiritual organization which is able to resist,
and must, as a matter of life and death, resist, the
progress of science and modern civilization, manages
her affairs much better.

It was my fortune some time ago to pay a visit to
one of the most important of the institutions in which
the clergy of the Roman Catholic Church in these islands
are trained; and it seemed to me that the difference
between these men and the comfortable champions of
Anglicanism and of Dissent, was comparable to the
difference between our gallant Volunteers and the
trained veterans of Napoleon's Old Guard.

The Catholic priest is trained to know his business,
and do it effectually. The professors of the college in
question, learned, zealous, and determined men, permitted me to speak frankly with them. We talked like
outposts of opposed armies during a truce—as friendly
enemies; and when I ventured to point out the difficulties their students would have to encounter from
scientific thought, they replied : "Our Church has lasted
many ages, and has passed safely through many storms.
The present is but a new gust of the old tempest, and
we do not turn out our young men less fitted to weather
it, than they have been, in former times, to cope with
the difficulties of those times. The heresies of the day
are explained to them by their professors of philosophy
and science, and they are taught how those heresies are
to be met."

I heartily respect an organization which faces its enemies in this way; and I wish that all ecclesiastical organizations were in as effective a condition. I think it would be better, not only for them, but for us. The army of liberal thought is, at present, in very loose order; and many a spirited free-thinker makes use of his freedom mainly to vent nonsense. We should be the better for a vigorous and watchful enemy to hammer us into cohesion and discipline; and I, for one, lament that the bench of Bishops cannot show a man of the calibre of Butler of the "Analogy," who, if he were alive, would make short work of much of the current *à priori* "infidelity."

I hope you will consider that the arguments I have now stated, even if there were no better ones, constitute a sufficient apology for urging the introduction of science into schools. The next question to which I have to address myself is, What sciences ought to be thus taught? And this is one of the most important of questions, because my side (I am afraid I am a terribly candid friend) sometimes spoils its cause by going in for too much. There are other forms of culture beside physical science; and I should be profoundly sorry to see the fact forgotten, or even to observe a tendency to starve, or cripple, literary, or æsthetic, culture for the sake of science. Such a narrow view of the nature of education has nothing to do with my firm conviction that a complete and thorough scientific culture ought to be introduced into all schools. By this, however, I do not mean that every schoolboy should be taught everything in science. That would be a very absurd thing to conceive, and a very mischievous thing to attempt. What I mean is, that no boy nor girl should leave school without possessing a grasp of the general character of science, and without having been disciplined, more or less, in the methods of all sciences; so that, when turned into the world to make their own way, they shall be prepared to face scientific problems, not by knowing at once the conditions of every problem, or by being able at once to solve it; but by being familiar with the general current of scientific thought, and by being able to apply the methods of science in the proper way, when they have acquainted themselves with the conditions of the special problem.

That is what I understand by scientific education. To furnish a boy with such an education, it is by no means necessary that he should devote his whole school existence to physical science: in fact, no one would lament so one-sided a proceeding more than I. Nay more, it is not necessary for him to give up more than a moderate share of his time to such studies, if they be properly selected and arranged, and if he be trained in them in a fitting manner.

I conceive the proper course to be somewhat as follows. To begin with, let every child be instructed in those general views of the phenomena of Nature for which we have no exact English name. The nearest approximation to a name for what I mean, which we possess, is " physical geography." The Germans have a better, " Erdkunde," (" earth knowledge " or " geology " in its etymological sense,) that is to say, a general knowledge of the earth, and what is on it, in it, and about it. If any one who has had experience of the ways of young children will call to mind their questions, he will find that so far as they can be put into any scientific category, they come under this head of " Erdkunde." The child asks, " What is the moon, and why does it shine ? " " What is this water, and where does it run ? " " What is the wind ? " " What makes the waves in the sea ? " " Where does this animal live, and what is the use of that plant ? " And if not snubbed and stunted by being told not to ask foolish questions, there is no limit to the intellectual craving of a young child ; nor any bounds to the slow, but solid, accretion of knowledge and development of the thinking faculty in this way. To all such questions, answers which are necessarily incomplete, though true as far as they go, may be given by any teacher whose ideas represent real knowledge and not mere book learning ; and a panoramic view of Nature, accompanied by a strong infusion of the scientific habit of mind, may thus be placed within the reach of every child of nine or ten.

After this preliminary opening of the eyes to the great spectacle of the daily progress of Nature, as the reasoning faculties of the child grow, and he becomes familiar with the use of the tools of knowledge—reading, writing, and elementary mathematics—he should pass on to what is, in the more strict sense, physical science. Now there are two kinds of physical science : the one regards form and the relation of forms to one another ; the other deals with causes and effects. In many of what we term our sciences, these two kinds are mixed up together ; but systematic botany is a pure example of the former kind, and physics of the latter kind, of science. Every educational advantage which training in physical science can give is obtainable from the proper study of these two; and I should be contented, for the present, if they, added to our " Erdkunde," furnished the whole of the scientific curriculum of schools. Indeed, I conceive it would be one of the greatest boons which could be conferred upon England, if henceforward every child in the country were instructed in the general knowledge of the things about it, in the elements of physics, and of botany. But I should be still better pleased if there could be added somewhat of chemistry, and an elementary acquaintance with human physiology.

So far as school education is concerned, I want to go no further just now; and I believe that such instruction would make an excellent introduction to that preparatory scientific training which, as I have indicated, is so essential for the successful pursuit of our most important professions. But this modicum of instruction must be so given as to ensure real knowledge and practical discipline. If scientific education is to be dealt with as mere book-work, it will be better not to attempt it, but to stick to the Latin Grammar, which makes no pretence to be anything but bookwork.

If the great benefits of scientific training are sought, it is essential that such training should be real: that is to say, that the mind of the scholar should be brought into direct relation with fact, that he should not merely be told a thing, but made to see by the use of his own intellect and ability that the thing is *so* and no otherwise. The great peculiarity of scientific training, that in virtue of which it cannot be replaced by any other discipline whatsoever, is this bringing of the mind directly into contact with fact, and practising the intellect in the completest form of induction; that is to say, in drawing conclusions from particular facts made known by immediate observation of Nature.

The other studies which enter into ordinary education do not discipline the mind in this way. Mathematical training is almost purely deductive. The mathematician starts with a few simple propositions, the proof of which is so obvious that they are called self-evident, and the rest of his work consists of subtle deductions from them. The teaching of languages, at any rate as ordinarily practised, is of the same general nature,—authority and tradition furnish the data, and the mental operations of the scholar are deductive.

Again: if history be the subject of study, the facts are still taken upon the evidence of tradition and authority. You cannot make a boy see the battle of Thermopylæ for himself, or know, of his own knowledge, that Cromwell once ruled England. There is no getting into direct contact with natural fact by this road; there is no dispensing with authority, but rather a resting upon it.

In all these respects, science differs from other educational discipline, and prepares the scholar for common life. What have we to do in every-day life? Most of the business which demands our attention is matter of fact, which needs, in the first place, to be accurately observed or apprehended; in the second, to be interpreted by inductive and deductive reasonings, which are altogether similar in their nature to those employed in science. In the one case, as in the other, whatever is taken for granted is so taken at one's own peril; fact

and reason are the ultimate arbiters, and patience and
honesty are the great helpers out of difficulty.

But if scientific training is to yield its most eminent
results, it must, I repeat, be made practical.   That is to
say, in explaining to a child the general phenomena of
Nature, you must, as far as possible, give reality to your
teaching by object-lessons; in teaching him botany, he
must handle the plants and dissect the flowers for himself; in teaching him physics and chemistry, you must
not be solicitous to fill him with information, but you
must be careful that what he learns he knows of his own
knowledge.   Don't be satisfied with telling him that a
magnet attracts iron.   Let him see that it does; let him
feel the pull of the one upon the other for himself.   And,
especially, tell him that it is his duty to doubt until he
is compelled, by the absolute authority of Nature, to
believe that which is written in books.   Pursue this
discipline carefully and conscientiously, and you may
make sure that, however scanty may be the measure of
information which you have poured into the boy's mind,
you have created an intellectual habit of priceless value
in practical life.

One is constantly asked, When should this scientific
education be commenced?   I should say with the dawn
of intelligence.   As I have already said, a child seeks
for information about matters of physical science as soon
as it begins to talk.   The first teaching it wants is an
object-lesson of one sort or another; and as soon as it
is fit for systematic instruction of any kind, it is fit
for a modicum of science.

People talk of the difficulty of teaching young
children such matters, and in the same breath insist

upon their learning their Catechism, which contains propositions far harder to comprehend than anything in the educational course I have proposed. Again, I am incessantly told that we, who advocate the introduction of science into schools, make no allowance for the stupidity of the average boy or girl; but, in my belief, that stupidity, in nine cases out of ten, "*fit, non nascitur,*" and is developed by a long process of parental and pedagogic repression of the natural intellectual appetites, accompanied by a persistent attempt to create artificial ones for food which is not only tasteless, but essentially indigestible.

Those who urge the difficulty of instructing young people in science are apt to forget another very important condition of success—important in all kinds of teaching, but most essential, I am disposed to think, when the scholars are very young. This condition is, that the teacher should himself really and practically know his subject. If he does, he will be able to speak of it in the easy language, and with the completeness of conviction, with which he talks of any ordinary every-day matter. If he does not, he will be afraid to wander beyond the limits of the technical phraseology which he has got up; and a dead dogmatism, which oppresses, or raises opposition, will take the place of the lively confidence, born of personal conviction, which cheers and encourages the eminently sympathetic mind of childhood.

I have already hinted that such scientific training as we seek for may be given without making any extravagant claim upon the time now devoted to education. We ask only for " a most favoured nation " clause in our treaty with the schoolmaster ; we demand no more than that science shall have as much time given to it as any other single subject—say four hours a week in each class of an ordinary school.

For the present, I think men of science would be well content with such an arrangement as this ; but, speaking for myself, I do not pretend to believe that such an arrangement can be, or will be, permanent. In these times the educational tree seems to me to have its roots in the air, its leaves and flowers in the ground ; and, I confess, I should very much like to turn it upside down, so that its roots might be solidly embedded among the facts of Nature, and draw thence a sound nutriment for the foliage and fruit of literature and of art. No educational system can have a claim to permanence, unless it recognises the truth that education has two great ends to which everything else must be subordinated. The one of these is to increase knowledge ; the other is to develop the love of right and the hatred of wrong.

With wisdom and uprightness a nation can make its way worthily, and beauty will follow in the footsteps of the two, even if she be not specially invited ; while there is perhaps no sight in the whole world more saddening and revolting than is offered by men sunk in ignorance of everything but what other men have written ; seemingly devoid of moral belief or guidance ; but with the sense of beauty so keen, and the power of expression so cultivated, that their sensual caterwauling may be almost mistaken for the music of the spheres.

At present, education is almost entirely devoted to the cultivation of the power of expression, and of the sense of literary beauty. The matter of having anything to say, beyond a hash of other people's opinions, or of possessing any criterion of beauty, so that we may distinguish between the Godlike and the devilish, is left aside as of no moment. I think I do not err in saying that if science were made the foundation of education, instead of being, at most, stuck on as cornice to the edifice, this state of things could not exist.

In advocating the introduction of physical science as a leading element in education, I by no means refer only to the higher schools. On the contrary, I believe that such a change is even more imperatively called for in those primary schools, in which the children of the poor are expected to turn to the best account the little time they can devote to the acquisition of knowledge. A great step in this direction has already been made by the establishment of science-classes under the Department of Science and Art,—a measure which came into existence unnoticed, but which will, I believe, turn out to be of more importance to the welfare of the people, than many political changes, over which the noise of battle has rent the air.

Under the regulations to which I refer, a schoolmaster can set up a class in one or more branches of science; his pupils will be examined, and the State will pay him, at a certain rate, for all who succeed in passing. I have acted as an examiner under this system from the beginning of its establishment, and this year I expect to have not fewer than a couple of thousand sets of answers to questions in Physiology, mainly from young people of the artisan class, who have been taught in the schools which are now scattered all over Great Britain and Ireland. Some of my colleagues, who have to deal with subjects such as Geometry, for which the present teaching power is better organized, I understand are likely to have three or four times as many papers.   So far as my own subjects are concerned, I can undertake to say that a great deal of the teaching, the results of which are before me in these examinations, is very sound and good; and I think it is in the power of the examiners, not only to keep up the present standard, but to cause an almost unlimited improvement.   Now what does this mean ?   It means that by holding out a very moderate inducement, the masters of primary schools in many parts of the country have been led to convert them into little foci of scientific instruction; and that they and their pupils have contrived to find, or to make, time enough to carry out this object with a very considerable degree of efficiency.   That efficiency will, I doubt not, be very much increased as the system becomes known and perfected, even with the very limited leisure left to masters and teachers on week-days.   And this leads me to ask, Why should scientific teaching be limited to week-days ?

Ecclesiastically-minded persons are in the habit of calling things they do not like by very hard names, and I should not wonder if they brand the proposition I am about to make as blasphemous, and worse.   But, not minding this, I venture to ask, Would there really be anything wrong in using part of Sunday for the purpose of instructing those who have no other leisure, in a knowledge of the phenomena of Nature, and of man's relation to nature ?

I should like to see a scientific Sunday-school in every parish, not for the purpose of superseding any existing

means of teaching the people the things that are for
their good, but side by side with them. I cannot but
think that there is room for all of us to work in help-
ing to bridge over the great abyss of ignorance which
lies at our feet.

And if any of the ecclesiastical persons to whom I
have referred, object that they find it derogatory to the
honour of the God whom they worship, to awaken the
minds of the young to the infinite wonder and majesty
of the works which they proclaim His, and to teach
them those laws which must needs be His laws, and
therefore of all things needful for man to know—I can
only recommend them to be let blood and put on low
diet. There must be something very wrong going on
in the instrument of logic, if it turns out such conclusions from such premisses.

ON THE EDUCATIONAL VALUE OF THE NATURAL HISTORY SCIENCES.

THE subject to which I have to beg your attention during the ensuing hour is "The Relation of Physiological Science to other branches of Knowledge."

Had circumstances permitted of the delivery, in their strict logical order, of that series of discourses of which the present lecture is a member, I should have preceded my friend and colleague Mr. Henfrey, who addressed you on Monday last; but while, for the sake of that order, I must beg you to suppose that this discussion of the Educational bearings of Biology in general *does* precede that of Special Zoology and Botany, I am rejoiced to be able to take advantage of the light thus already thrown upon the tendency and methods of Physiological Science.

Regarding Physiological Science, then, in its widest sense—as the equivalent of *Biology*—the Science of Individual Life—we have to consider in succession:

1. Its position and scope as a branch of knowledge.
2. Its value as a means of mental discipline.

3. Its worth as practical information.

And lastly,

4. At what period it may best be made a branch of Education.

Our conclusions on the first of these heads must depend, of course, upon the nature of the subject-matter of Biology; and I think a few preliminary considerations will place before you in a clear light the vast difference which exists between the living bodies with which Physiological science is concerned, and the remainder of the universe;—between the phæno-mena of Number and Space, of Physical and of Chemical force, on the one hand, and those of Life on the other.

The mathematician, the physicist, and the chemist contemplate things in a condition of rest; they look upon a state of equilibrium as that to which all bodies normally tend.

The mathematician does not suppose that a quantity will alter, or that a given point in space will change its direction with regard to another point, sponta-neously. And it is the same with the physicist. When Newton saw the apple fall, he concluded at once that the act of falling was not the result of any power inherent in the apple, but that it was the result of the action of something else on the apple. In a similar manner, all physical force is regarded as the disturbance of an equilibrium to which things tended before its exertion,—to which they will tend again after its cessation.

The chemist equally regards chemical change in a body, as the effect of the action of something external to the body changed. A chemical compound once formed would persist for ever, if no alteration took place in surrounding conditions.

But to the student of Life the aspect of nature is reversed. Here, incessant, and, so far as we know, spontaneous change is the rule, rest the exception—the anomaly to be accounted for. Living things have no inertia, and tend to no equilibrium.

Permit me, however, to give more force and clearness to these somewhat abstract considerations, by an illustration or two.

Imagine a vessel full of water, at the ordinary temperature, in an atmosphere saturated with vapour. The *quantity* and the *figure* of that water will not change, so far as we know, for ever.

Suppose a lump of gold be thrown into the vessel—motion and disturbance of figure exactly proportional to the momentum of the gold will take place. But after a time the effects of this disturbance will subside—equilibrium will be restored, and the water will return to its passive state.

Expose the water to cold—it will solidify—and in so doing its particles will arrange themselves in definite crystalline shapes. But once formed, these crystals change no further.

Again, substitute for the lump of gold some substance capable of entering into chemical relations with the water :—say, a mass of that substance which is called " protein "—the substance of flesh :—a very considerable disturbance of equilibrium will take place—all sorts of chemical compositions and decompositions will occur ; but in the end, as before, the result will be the resump tion of a condition of rest.

Instead of such a mass of *dead* protein, however, take a particle of *living* protein—one of those minute microscopic living things which throng our pools, and are known as Infusoria—such a creature, for instance, as an Euglena, and place it in our vessel of water. It is a round mass provided with a long filament, and except in this peculiarity of shape, presents no appreciable physical or chemical difference whereby it might be distinguished from the particle of dead protein.

But the difference in the phænomena to which it will give rise is immense: in the first place it will develop a vast quantity of physical force—cleaving the water in all directions with considerable rapidity by means of the vibrations of the long filament or cilium.

Nor is the amount of chemical energy which the little creature possesses less striking. It is a perfect laboratory in itself, and it will act and react upon the water and the matters contained therein ; converting them into new compounds resembling its own substance, and, at the same time, giving up portions of its own substance which have become effete.

Furthermore, the Euglena will increase in size ; but this increase is by no means unlimited, as the increase of a crystal might be. After it has grown to a certain extent it divides, and each portion assumes the form of the original, and proceeds to repeat the process of growth and division.

Nor is this all. For after a series of such divisions and subdivisions, these minute points assume a totally new form, lose their long tails—round themselves, and secrete a sort of envelope or box, in which they remain shut up for a time, eventually to resume, directly or indirectly, their primitive mode of existence.

Now, so far as we know, there is no natural limit to the existence of the Euglena, or of any other living germ. A living species once launched into existence tends to live for ever.

Consider how widely different this living particle is from the dead atoms with which the physicist and chemist have to do!

The particle of gold falls to the bottom and rests— the particle of dead protein decomposes and disappears —it also rests: but the *living* protein mass neither tends to exhaustion of its forces nor to any permanency of form, but is essentially distinguished as a disturber of equilibrium so far as force is concerned,—as under-going continual metamorphosis and change, in point of form.

Tendency to equilibrium of force and to permanency of form then, are the characters of that portion of the universe which does not live—the domain of the chemist and physicist.

Tendency to disturb existing equilibrium,—to take on forms which succeed one another in definite cycles, is the character of the living world.

What is the cause of this wonderful difference between the dead particle and the living particle of matter appearing in other respects identical? that difference to which we give the name of Life?

I, for one, cannot tell you. It may be that, by and by, philosophers will discover some higher laws of which the facts of life are particular cases—very possibly they will find out some bond between physico-chemical phænomena on the one hand, and vital phænomena on the other. At present, however, we assuredly know of none ; and I think we shall exercise a wise humility in confessing that, for us at least, this successive assumption of different states—(external conditions remaining the same)—this *spontaneity of action*—if I may use a term which implies more than I would be answerable for—which constitutes so vast and plain a practical distinction between living bodies and those which do not live, is an ultimate fact ; indicating as such, the existence of a broad line of demarcation between the subject-matter of Biological and that of all other sciences,

For I would have it understood that this simple Euglena is the type of *all* living things, so far as the distinction between these and inert matter is concerned. That cycle of changes, which is constituted by perhaps not more than two or three steps in the Euglena, is as clearly manifested in the multitudinous stages through which the germ of an oak or of a man passes. Whatever forms the Living Being may take on, whether simple or complex, *production, growth, reproduction*, are the phænomena which distinguish it from that which does not live.

If this be true, it is clear that the student, in passing from the physico-chemical to the physiological sciences, enters upon a totally new order of facts ; and it will next be for us to consider how far these new facts involve *new* methods, or require a modification of those with which he is already acquainted. Now a great deal is said about the peculiarity of the scientific method in general, and of the different methods which are pursued in the different sciences. The Mathematics are said to have one special method; Physics another, Biology a third, and so forth. For my own part, I must confess that I do not understand this phraseology.

So far as I can arrive at any clear comprehension of the matter, Science is not, as many would seem to suppose, a modification of the black art, suited to the tastes of the nineteenth century, and flourishing mainly in consequence of the decay of the Inquisition.

Science is, I believe, nothing but *trained and organized common sense*, differing from the latter only as a veteran may differ from a raw recruit: and its methods differ from those of common sense only so far as the guardsman's cut and thrust differ from the manner in which a savage wields his club. The primary power is the same in each case, and perhaps the untutored savage has the more brawny arm of the two. The *real* advantage lies in the point and polish of the swordsman's weapon; in the trained eye quick to spy out the weakness of the adversary; in the ready hand prompt to follow it on the instant. But after all, the sword exercise is only the hewing and poking of the clubman developed and perfected.

So, the vast results obtained by Science are won by no mystical faculties, by no mental processes, other than those which are practised by every one of us, in the humblest and meanest affairs of life. A detective policeman discovers a burglar from the marks made by his shoe, by a mental process identical with that by which Cuvier restored the extinct animals of Montmartre from fragments of their bones. Nor does that process of induction and deduction by which a lady, finding a stain of a peculiar kind upon her dress, concludes that somebody has upset the inkstand thereon, differ in any way, in kind, from that by which Adams and Leverrier discovered a new planet.

The man of science, in fact, simply uses with scrupulous exactness, the methods which we all, habitually and at every moment, use carelessly; and the man of business must as much avail himself of the scientific method—must be as truly a man of science—as the veriest bookworm of us all; though I have no doubt that the man of business will find himself out to be a philosopher with as much surprise as M. Jourdain exhibited, when he discovered that he had been all his life talking prose. If, however, there be no real difference between the methods of science and those of common life, it would seem, on the face of the matter, highly improbable that there should be any difference between the methods of the different sciences; nevertheless, it is constantly taken for granted, that there is a very wide difference between the Physiological and other sciences in point of method.

In the first place it is said—and I take this point first, because the imputation is too frequently admitted by Physiologists themselves—that Biology differs from the Physico-chemical and Mathematical sciences in being "inexact."

Now, this phrase "inexact" must refer either to the *methods* or to the *results* of Physiological science.

It cannot be correct to apply it to the methods; for, as I hope to show you by and by, these are identical in all sciences, and whatever is true of Physiological method is true of Physical and Mathematical method.

Is it then the *results* of Biological science which are

"inexact"? I think not. If I say that respiration is performed by the lungs; that digestion is effected in the stomach; that the eye is the organ of sight; that the jaws of a vertebrated animal never open sideways, but always up and down; while those of an annulose animal always open sideways, and never up and down—I am enumerating propositions which are as exact as anything in Euclid. How then has this notion of the inexactness of Biological science come about? I believe from two causes: first, because, in consequence of the great complexity of the science and the multitude of interfering conditions, we are very often only enabled to predict approximatively what will occur under given circumstances; and secondly, because, on account of the comparative youth of the Physiological sciences, a great many of their laws are still imperfectly worked out. But, in an educational point of view, it is most important to distinguish between the essence of a science and the accidents which surround it; and essentially, the methods and results of Physiology are as exact as those of Physics or Mathematics.

It is said that the Physiological method is especially *comparative* [1]; and this dictum also finds favour in the

---

[1] "In the third place, we have to review the method of Comparison, which is so specially adapted to the study of living bodies, and by which, above all others, that study must be advanced. In Astronomy, this method is necessarily inapplicable; and it is not till we arrive at Chemistry that this third means of investigation can be used, and then only in subordination to the two others. It is in the study, both statical and dynamical, of living bodies that it first acquires its full development; and its use elsewhere can be only through its application here."—COMTE'S *Positive Philosophy*, translated by Miss Martineau. Vol. i. p. 372.

By what method does M. Comte suppose that the equality or inequality of forces and quantities and the dissimilarity or similarity of forms—points of some slight importance not only in Astronomy and Physics, but even in Mathematics—are ascertained, if not by Comparison?

eyes of many. I should be sorry to suggest that the speculators on scientific classification have been misled by the accident of the name of one leading branch of Biology—*Comparative Anatomy ;* but I would ask whether *comparison,* and that classification which is the result of comparison, are not the essence of every science whatsoever? How is it possible to discover a relation of cause and effect of *any* kind without comparing a series of cases together in which the supposed cause and effect occur singly, or combined? So far from comparison being in any way peculiar to Biological science, it is, I think, the essence of every science.

A speculative philosopher again tells us that the Biological sciences are distinguished by being sciences of observation and not of experiment![1]

Of all the strange assertions into which speculation without practical acquaintance with a subject may lead even an able man, I think this is the very strangest. Physiology not an experimental science! Why, there is not a function of a single organ in the body which has not been determined wholly and solely by experiment. How did Harvey determine the nature of the circulation, except by experiment? How did Sir Charles Bell determine the functions of the roots of the spinal nerves,

---

[1] " Proceeding to the second class of means,—Experiment cannot but be less and less decisive, in proportion to the complexity of the phænomena to be explored ; and therefore we saw this resource to be less effectual in chemistry than in physics : and we now find that it is eminently useful in chemistry in comparison with physiology. *In fact, the nature of the phænomena seems to offer almost insurmountable impedients to any extensive and prolific application of such a procedure in biology.*"—Comte, vol i. p. 367.

M. Comte, as his manner is, contradicts himself two pages further on, but that will hardly relieve him from the responsibility of such a paragraph as the above.

save by experiment? How do we know the use of a
nerve at all, except by experiment? Nay, how do you
know even that your eye is your seeing apparatus, unless
you make the experiment of shutting it; or that your
ear is your hearing apparatus, unless you close it up and
thereby discover that you become deaf?

It would really be much more true to say that Physiology is *the* experimental science *par excellence* of all
sciences; that in which there is least to be learnt by
mere observation, and that which affords the greatest
field for the exercise of those faculties which characterise
the experimental philosopher. I confess, if any one
were to ask me for a model application of the logic of
experiment, I should know no better work to put into
his hands than Bernard's late Researches on the Functions of the Liver.[1]

Not to give this lecture a too controversial tone, however, I must only advert to one more doctrine, held by a
thinker of our own age and country, whose opinions are
worthy of all respect. It is, that the Biological sciences
differ from all others, inasmuch as in *them* classification
takes place by type and not by definition.[2]

---

[1] "Nouvelle Fonction du Foie considéré comme organe producteur de
matière sucrée chez l'Homme et les Animaux," par M. Claude Bernard.

[2] "*Natural Groups given by Type, not by Definition* .... The class is
steadily fixed, though not precisely limited ; it is given, though not circumscribed ; it is determined, not by a boundary-line without, but by a central
point within ; not by what it strictly excludes, but what it eminently includes ;
by an example, not by a precept ; in short, instead of Definition we have a
Type for our director. A type is an example of any class, for instance, a
species of a genus, which is considered as eminently possessing the characters
of the class. All the species which have a greater affinity with this type-
species than with any others, form the genus, and are ranged about it,
deviating from it in various directions and different degrees."—WHEWELL,
The Philosophy of the Inductive Sciences, vol. i. pp. 476, 477.

It is said, in short, that a natural-history class is not capable of being defined—that the class Rosaceæ, for instance, or the class of Fishes, is not accurately and absolutely definable, inasmuch as its members will present exceptions to every possible definition; and that the members of the class are united together only by the circumstance that they are all more like some imaginary average rose or average fish, than they resemble anything else.

But here, as before, I think the distinction has arisen entirely from confusing a transitory imperfection with an essential character. So long as our information concerning them is imperfect, we class all objects together according to resemblances which we *feel*, but cannot *define*: we group them round *types*, in short. Thus, if you ask an ordinary person what kinds of animals there are, he will probably say, beasts, birds, reptiles, fishes, insects, &c. Ask him to define a beast from a reptile, and he cannot do it; but he says, things like a cow or a horse are beasts, and things like a frog or a lizard are reptiles. You see *he does* class by type, and not by definition. But how does this classification differ from that of the scientific Zoologist? How does the meaning of the scientific class-name of " Mammalia " differ from the unscientific of " Beasts "?

Why, exactly because the former depends on a definition, the latter on a type. The class Mammalia is scientifically defined as " all animals which have a vertebrated skeleton and suckle their young." Here is no reference to type, but a definition rigorous enough for a geometrician. And such is the character which every scientific naturalist recognises as that to which his classes must aspire—knowing, as he does, that classification by type is simply an acknowledgment of ignorance and a temporary device.

So much in the way of negative argument as against the reputed differences, between Biological and other methods. No such differences, I believe, really exist. The subject-matter of Biological science is different from that of other sciences, but the methods of all are identical; and these methods are—

1. *Observation* of facts—including under this head that *artificial observation* which is called *experiment.*

2. That process of tying up similar facts into bundles, ticketed and ready for use, which is called *Comparison* and *Classification,*—the results of the process, the ticketed bundles, being named *General propositions.*

3. *Deduction,* which takes us from the general proposition to facts again—teaches us, if I may so say, to anticipate from the ticket what is inside the bundle. And finally—

4. *Verification,* which is the process of ascertaining whether, in point of fact, our anticipation is a correct one.

Such are the methods of all science whatsoever; but perhaps you will permit me to give you an illustration of their employment in the science of Life; and I will take as a special case, the establishment of the doctrine of the *Circulation of the Blood.*

In this case, *simple observation* yields us a knowledge of the existence of the blood from some accidental hæmorrhage, we will say : we may even grant that it informs us of the localization of this blood in particular vessels, the heart, &c., from some accidental cut or the like.   It teaches also the existence of a pulse in various parts of the body, and acquaints us with the structure of the heart and vessels.

Here, however, *simple observation* stops, and we must have recourse to *experiment*.

You tie a vein, and you find that the blood accumulates on the side of the ligature opposite the heart.   You tie an artery, and you find that the blood accumulates on the side near the heart.   Open the chest, and you see the heart contracting with great force.   Make openings into its principal cavities, and you will find that all the blood flows out, and no more pressure is exerted on either side of the arterial or venous ligature.

Now all these facts, taken together, constitute the evidence that the blood is propelled by the heart through the arteries, and returns by the veins—that, in short, the blood circulates.

Suppose our experiments and observations have been made on horses, then we group and ticket them into a general proposition, thus :—*all horses have a circulation of their blood.*

Henceforward a horse is a sort of indication or label, telling us where we shall find a peculiar series of phænomena called the circulation of the blood.

Here is our *general proposition* then.

How and when are we justified in making our next step—a *deduction* from it ?

Suppose our physiologist, whose experience is limited to horses, meets with a zebra for the first time,—will he suppose that this generalization holds good for zebras also ?

That depends very much on his turn of mind.   But

we will suppose him to be a bold man. He will say, " The zebra is certainly not a horse, but it is very like one,—so like, that it must be the ' ticket' or mark of a blood-circulation also ; and, I conclude that the zebra has a circulation."

That is a deduction, a very fair deduction, but by no means to be considered scientifically secure. This last quality in fact can only be given by *verification*—that is, by making a zebra the subject of all the experiments performed on the horse. Of course, in the present case, the *deduction* would be *confirmed* by this process of verification, and the result would be, not merely a positive widening of knowledge, but a fair increase of confidence in the truth of one's generalizations in other cases.

Thus, having settled the point in the zebra and horse, our philosopher would have great confidence in the existence of a circulation in the ass. Nay, I fancy most persons would excuse him, if in this case he did not take the trouble to go through the process of verification at all ; and it would not be without a parallel in the history of the human mind, if our imaginary physiologist now maintained that he was acquainted with asinine circulation *à priori*.

However, if I might impress any caution upon your minds, it is, the utterly conditional nature of all our knowledge,—the danger of neglecting the process of verification under any circumstances ; and the film upon which we rest, the moment our deductions carry us beyond the reach of this great process of verification. There is no better instance of this than is afforded by the history of our knowledge of the circulation of the blood in the animal kingdom until the year 1824. In every animal possessing a circulation at all, which had been observed up to that time, the current of the blood was known to take one definite and invariable direction. Now, there is a class of animals called *Ascidians*, which possess a heart and a circulation, and up to the period of which I speak, no one would have dreamt of questioning the propriety of the deduction, that these creatures have a circulation in one direction ; nor would any one have thought it worth while to verify the point. But, in that year, M. von Hasselt happening to examine a transparent animal of this class, found to his infinite surprise, that after the heart had beat a certain number of times, it stopped, and then began beating the opposite way—so as to reverse the course of the current, which returned by and by to its original direction.

I have myself timed the heart of these little animals. I found it as regular as possible in its periods of reversal : and I know no spectacle in the animal kingdom more wonderful than that which it presents—all the more wonderful that to this day it remains an unique fact, peculiar to this class among the whole animated world. At the same time I know of no more striking case of the necessity of the *verification* of even those deductions which seem founded on the widest and safest inductions.

Such are the methods of Biology—methods which are obviously identical with those of all other sciences, and therefore wholly incompetent to form the ground of any distinction between it and them.[1]

---

[1] Save for the pleasure of doing so, I need hardly point out my obligations to Mr. J. S. Mill's " System of Logic," in this view of scientific method.

But I shall be asked at once, Do you mean to say that there is no difference between the habit of mind of a mathematician and that of a naturalist? Do you imagine that Laplace might have been put into the Jardin des Plantes, and Cuvier into the Observatory, with equal advantage to the progress of the sciences they professed?

To which I would reply, that nothing could be further from my thoughts. But different habits and various special tendencies of two sciences do not imply different methods. The mountaineer and the man of the plains have very different habits of progression, and each would be at a loss in the other's place; but the method of progression, by putting one leg before the other, is the same in each case. Every step of each is a combination of a lift and a push; but the mountaineer lifts more and the lowlander pushes more. And I think the case of two sciences resembles this.

I do not question for a moment, that while the Mathematician is busied with deductions *from* general propositions, the Biologist is more especially occupied with observation, comparison, and those processes which lead *to* general propositions. All I wish to insist upon is, that this difference depends not on any fundamental distinction in the sciences themselves, but on the accidents of their subject-matter, of their relative complexity, and consequent relative perfection.

The Mathematician deals with two properties of objects only, number and extension, and all the inductions he wants have been formed and finished ages ago. He is occupied now with nothing but deduction and verification.

The Biologist deals with a vast number of properties of objects, and his inductions will not be completed, I fear, for ages to come ; but when they are, his science will be as deductive and as exact as the Mathematics themselves.

Such is the relation of Biology to those sciences which deal with objects having fewer properties than itself. But as the student, in reaching Biology, looks back upon sciences of a less complex and therefore more perfect nature ; so, on the other hand, does he look forward to other more complex and less perfect branches of knowledge. Biology deals only with living beings as isolated things—treats only of the life of the individual : but there is a higher division of science still, which considers living beings as aggregates—which deals with the relation of living beings one to another—the science which *observes* men—whose *experiments* are made by nations one upon another, in battle-fields—whose *general propositions* are embodied in history, morality, and religion— whose *deductions* lead to our happiness or our misery, —and whose *verifications* so often come too late, and serve only

" To point a moral or adorn a tale "—

I mean the science of Society or *Sociology*.

I think it is one of the grandest features of Biology, that it occupies this central position in human knowledge. There is no side of the human mind which physiological study leaves uncultivated. Connected by innumerable ties with abstract science, Physiology is yet in the most intimate relation with humanity ; and by teaching us that law and order, and a definite scheme of development, regulate even the strangest and wildest manifestations of individual life, she prepares the student to look for a goal even amidst the erratic wanderings of mankind, and to believe that history offers something more than an entertaining chaos—a journal of a toilsome, tragi-comic march nowhither.

The preceding considerations have, I hope, served to indicate the replies which befit the two first of the questions which I set before you at starting, viz. what is the range and position of Physiological Science as a branch of knowledge, and what is its value as a means of mental discipline.

Its *subject-matter* is a large moiety of the universe— its *position* is midway between the physico-chemical and the social sciences. Its *value* as a branch of discipline is partly that which it has in common with all sciences— the training and strengthening of common sense ; partly that which is more peculiar to itself—the great exercise which it affords to the faculties of observation and comparison ; and I may add, the *exactness* of knowledge which it requires on the part of those among its votaries who desire to extend its boundaries.

If what has been said as to the position and scope of Biology be correct, our third question—What is the practical value of physiological instruction ?—might, one would think, be left to answer itself.

On other grounds even, were mankind deserving of the title "rational," which they arrogate to themselves, there can be no question that they would consider, as the most necessary of all branches of instruction for themselves and for their children, that which professes to acquaint them with the conditions of the existence they prize so highly—which teaches them how to avoid disease and to cherish health, in themselves and those who are dear to them.

I am addressing, I imagine, an audience of educated persons ; and yet I dare venture to assert that, with the exception of those of my hearers who may chance to have received a medical education, there is not one who could tell me what is the meaning and use of an act which he performs a score of times every minute, and whose suspension would involve his immediate death ;— I mean the act of breathing—or who could state in precise terms why it is that a confined atmosphere is injurious to health.

The *practical value* of Physiological knowledge ! Why is it that educated men can be found to maintain that a slaughter-house in the midst of a great city is rather a good thing than otherwise ?—that mothers persist in exposing the largest possible amount of surface of their children to the cold, by the absurd style of dress they adopt, and then marvel at the peculiar dispensation of Providence, which removes their infants by bronchitis and gastric fever ? Why is it that quackery rides ram- pant over the land ; and that not long ago, one of the largest public rooms in this great city could be filled by an audience gravely listening to the reverend expositor of the doctrine—that the simple physiological phænomena known as spirit-rapping, table-turning, phreno-magnetism, and by I know not what other absurd and inappropriate names, are due to the direct and personal agency of Satan?

Why is all this, except from the utter ignorance as to the simplest laws of their own animal life, which prevails among even the most highly educated persons in this country ?

But there are other branches of Biological Science, besides Physiology proper, whose practical influence, though less obvious, is not, as I believe, less certain. I have heard educated men speak with an ill-disguised contempt of the studies of the naturalist, and ask, not without a shrug, " What is the use of knowing all about these miserable animals—what bearing has it on human life ? "

I will endeavour to answer that question. I take it that all will admit there is definite Government of this universe—that its pleasures and pains are not scattered at random, but are distributed in accordance with orderly and fixed laws, and that it is only in accordance with all we know of the rest of the world, that there should be an agreement between one portion of the sensitive creation and another in these matters.

Surely then it interests us to know the lot of other animal creatures—however far below us, they are still the sole created things which share with us the capability of pleasure and the susceptibility to pain.

I cannot but think that he who finds a certain pro-portion of pain and evil inseparably woven up in the life of the very worms, will bear his own share with more courage and submission ; and will, at any rate, view with suspicion those weakly amiable theories of the Divine government, which would have us believe pain to be an oversight and a mistake,—to be corrected by and by. On the other hand, the predominance of happiness among living things—their lavish beauty—the secret and wonderful harmony which pervades them all, from the highest to the lowest, are equally striking refutations of that modern Manichean doctrine, which exhibits the world as a slave-mill, worked with many tears, for mere utilitarian ends.

There is yet another way in which natural history may, I am convinced, take a profound hold upon practical life,—and that is, by its influence over our finer feelings, as the greatest of all sources of that pleasure which is derivable from beauty. I do not pretend that natural-history knowledge, as such, can increase our sense of the beautiful in natural objects. I do not suppose that the dead soul of Peter Bell, of whom the great poet of nature says,—

> A primrose by the river's brim,
> A yellow primrose was to him,—
> And it was nothing more,—

would have been a whit roused from its apathy, by the information that the primrose is a Dicotyledonous Exogen, with a monopetalous corolla and central placentation. But I advocate natural-history knowledge from this point of view, because it would lead us to *seek* the beauties of natural objects, instead of trusting to chance to force them on our attention. To a person uninstructed in natural history, his country, or sea-side, stroll is a walk through a gallery filled with wonderful works of art, nine-tenths of which have their faces turned to the wall. Teach him something of natural history, and you place in his hands a catalogue of those which are worth turning round. Surely our innocent pleasures are not so abundant in this life, that we can afford to despise this or any other source of them. We should fear being banished for our neglect to that limbo, where the great Florentine tells us are those who, during this life, " wept when they might be joyful."

But I shall be trespassing unwarrantably on your kindness, if I do not proceed at once to my last point—the time at which Physiological Science should first form a part of the Curriculum of Education.

The distinction between the teaching of the facts of a science as instruction, and the teaching it systematically as knowledge, has already been placed before you in a previous lecture : and it appears to me, that, as with other sciences, the *common facts* of Biology—the uses of parts of the body—the names and habits of the living creatures which surround us—may be taught with advantage to the youngest child. Indeed, the avidity of children for this kind of knowledge, and the comparative ease with which they retain it, is something quite marvellous. I doubt whether any toy would be so acceptable to young children as a vivarium, of the same kind as, but of course on a smaller scale than, those admirable devices in the Zoological Gardens.

On the other hand, systematic teaching in Biology cannot be attempted with success until the student has attained to a certain knowledge of physics and chemistry : for though the phænomena of life are dependent neither on physical nor on chemical, but on vital forces, yet they result in all sorts of physical and chemical changes, which can only be judged by their own laws.

And now to sum up in a few words the conclusions to which I hope you see reason to follow me.

Biology needs no apologist when she demands a place —and a prominent place—in any scheme of education worthy of the name. Leave out the Physiological sciences from your curriculum, and you launch the student into the world, undisciplined in that science whose subject-matter would best develop his powers of observation; ignorant of facts of the deepest importance for his own and others' welfare; blind to the richest sources of beauty in God's creation; and unprovided with that belief in a living law, and an order manifesting itself in and through endless change and variety, which might serve to check and moderate that phase of despair through which, if he take an earnest interest in social problems, he will assuredly sooner or later pass.

Finally, one word for myself. I have not hesitated to speak strongly where I have felt strongly; and I am but too conscious that the indicative and imperative moods have too often taken the place of the more becoming subjunctive and conditional. I feel, therefore, how necessary it is to beg you to forget the personality of him who has thus ventured to address you, and to consider only the truth or error in what has been said.

VI.

ON THE STUDY OF ZOOLOGY.

NATURAL HISTORY is the name familiarly applied to the study of the properties of such natural bodies as minerals, plants, and animals; the sciences which embody the knowledge man has acquired upon these subjects are commonly termed Natural Sciences, in contradistinction to other, so-called " physical," sciences; and those who devote themselves especially to the pursuit of such sciences have been, and are, commonly termed " Naturalists."

Linnæus was a naturalist in this wide sense, and his " Systema Naturæ " was a work upon natural history, in the broadest acceptation of the term; in it, that great methodizing spirit embodied all that was known in his time of the distinctive characters of minerals, animals, and plants. But the enormous stimulus which Linnæus gave to the investigation of nature soon rendered it impossible that any one man should write another " Systema Naturæ," and extremely difficult for any one to become a naturalist such as Linnæus was.

Great as have been the advances made by all the three

branches of science, of old included under the title of natural history, there can be no doubt that zoology and botany have grown in an enormously greater ratio than mineralogy; and hence, as I suppose, the name of " natural history " has gradually become more and more definitely attached to these prominent divisions of the subject, and by " naturalist" people have meant more and more distinctly to imply a student of the structure and functions of living beings.

However this may be, it is certain that the advance of knowledge has gradually widened the distance between mineralogy and its old associates, while it has drawn zoology and botany closer together; so that of late years it has been found convenient (and indeed necessary) to associate the sciences which deal with vitality and all its phenomena under the common head of " biology;" and the biologists have come to repudiate any blood-relationship with their foster-brothers, the mineralogists.

Certain broad laws have a general application throughout both the animal and the vegetable worlds, but the ground common to these kingdoms of nature is not of very wide extent, and the multiplicity of details is so great, that the student of living beings finds himself obliged to devote his attention exclusively either to the one or the other. If he elects to study plants, under any aspect, we know at once what to call him; he is a botanist, and his science is botany. But if the investigation of animal life be his choice, the name generally applied to him will vary, according to the kind of animals he studies, or the particular phenomena of animal life to which he confines his attention. If the study of man is his object, he is called an anatomist, or

a physiologist, or an ethnologist; but if he dissects
animals, or examines into the mode in which their functions are performed, he is a comparative anatomist or
comparative physiologist. If he turns his attention to
fossil animals, he is a palæontologist. If his mind is
more particularly directed to the description, specific
discrimination, classification, and distribution of animals,
he is termed a zoologist.

For the purposes of the present discourse, however, I
shall recognise none of these titles save the last, which I
shall employ as the equivalent of botanist, and I shall
use the term zoology as denoting the whole doctrine
of animal life, in contradistinction to botany, which
signifies the whole doctrine of vegetable life.

Employed in this sense, zoology, like botany, is divisible into three great but subordinate sciences, morphology, physiology, and distribution, each of which
may, to a very great extent, be studied independently
of the other.

Zoological morphology is the doctrine of animal form
or structure. Anatomy is one of its branches, development is another; while classification is the expression
of the relations which different animals bear to one
another, in respect of their anatomy and their development.

Zoological distribution is the study of animals in
relation to the terrestrial conditions which obtain now,
or have obtained at any previous epoch of the earth's
history.

Zoological physiology, lastly, is the doctrine of the
functions or actions of animals. It regards animal bodies
as machines impelled by certain forces, and performing

an amount of work, which can be expressed in terms of the ordinary forces of nature. The final object of physiology is to deduce the facts of morphology, on the one hand, and those of distribution on the other, from the laws of the molecular forces of matter.

Such is the scope of zoology. But if I were to content myself with the enunciation of these dry definitions, I should ill exemplify that method of teaching this branch of physical science, which it is my chief business tonight to recommend. Let us turn away then from abstract definitions. Let us take some concrete living thing, some animal, the commoner the better, and let us see how the application of common sense and common logic to the obvious facts it presents, inevitably leads us into all these branches of zoological science.

I have before me a lobster. When I examine it, what appears to be the most striking character it presents ? Why, I observe that this part which we call the tail of the lobster, is made up of six distinct hard rings and a seventh terminal piece. If I separate one of the middle rings, say the third, I find it carries upon its under surface a pair of limbs or appendages, each of which consists of a stalk and two terminal pieces. So that I can represent a transverse section of the ring and its appendages upon the diagram board in this way.

If I now take the fourth ring I find it has the same structure, and so have the fifth and the second ; so that, in each of these divisions of the tail, I find parts which correspond with one another, a ring and two appendages ; and in each appendage a stalk and two end pieces. These corresponding parts are called, in the technical language of anatomy, " homologous parts." The ring of the third division is the "homologue" of the ring of the fifth, the appendage of the former is the homologue of the appendage of the latter. And, as each division exhibits corresponding parts in corresponding places, we say that all the divisions are constructed upon the same plan. But now let us consider the sixth division. It is similar to, and yet different from, the others. The ring is essentially the same as in the other divisions; but the appendages look at first as if they were very different; and yet when we regard them closely, what do we find? A stalk and two terminal divisions, exactly as in the others, but the stalk is very short and very thick, the terminal divisions are very broad and flat, and one of them is divided into two pieces.

I may say, therefore, that the sixth segment is like the others in plan, but that it is modified in its details.

The first segment is like the others, so far as its ring is concerned, and though its appendages differ from any of those yet examined in the simplicity of their structure, parts corresponding with the stem and one of the divisions of the appendages of the other segments can be readily discerned in them.

Thus it appears that the lobster's tail is composed of a series of segments which are fundamentally similar, though each presents peculiar modifications of the plan common to all. But when I turn to the fore part of the body I see, at first, nothing but a great shield-like shell, called technically the "carapace," ending in front in a sharp spine, on either side of which are the curious compound eyes, set upon the ends of stout moveable stalks. Behind these, on the under side of the body, are two pairs of long feelers, or antennæ, followed by six pairs of jaws, folded against one another over the mouth, and five pairs of legs, the foremost of these being the great pinchers, or claws, of the lobster.

It looks, at first, a little hopeless to attempt to find in this complex mass a series of rings, each with its pair of appendages, such as I have shown you in the abdomen, and yet it is not difficult to demonstrate their existence. Strip off the legs, and you will find that each pair is attached to a very definite segment of the under wall of the body; but these segments, instead of being the lower parts of free rings, as in the tail, are such parts of rings which are all solidly united and bound together; and the like is true of the jaws, the feelers, and the eye-stalks, every pair of which is borne upon its own special segment. Thus the conclusion is gradually forced upon us, that the body of the lobster is composed of as many rings as there are pairs of appendages, namely, twenty in all, but that the six hindmost rings remain free and moveable, while the fourteen front rings become firmly soldered together, their backs forming one continuous shield—the carapace.

Unity of plan, diversity in execution, is the lesson taught by the study of the rings of the body, and the same instruction is given still more emphatically by the appendages. If I examine the outermost jaw I find it consists of three distinct portions, an inner, a middle, and an outer, mounted upon a common stem; and if I compare this jaw with the legs behind it, or the jaws in front of it, I find it quite easy to see, that, in the legs, it is the part of the appendage which corresponds with the inner division, which becomes modified into what we know familiarly as the "leg," while the middle division disappears, and the outer division is hidden under the carapace. Nor is it more difficult to discern that, in the appendages of the tail, the middle division appears again and the outer vanishes; while, on the other hand, in the foremost jaw, the so-called mandible, the inner division only is left; and, in the same way, the parts of the feelers and of the eye-stalks can be identified with those of the legs and jaws.

But whither does all this tend? To the very remarkable conclusion that a unity of plan, of the same kind as that discoverable in the tail or abdomen of the lobster, pervades the whole organization of its skeleton, so that I can return to the diagram representing any one of the rings of the tail, which I drew upon the board, and by adding a third division to each appendage, I can use it as a sort of scheme or plan of any ring of the body. I can give names to all the parts of that figure, and then if I take any segment of the body of the lobster, I can point out to you exactly, what modification the general plan has undergone in that particular segment; what part has remained moveable, and what has become fixed to another; what has been excessively developed and metamorphosed, and what has been suppressed.

But I imagine I hear the question, How is all this to be tested? No doubt it is a pretty and ingenious way of looking at the structure of any animal, but is it anything more? Does Nature acknowledge, in any deeper way, this unity of plan we seem to trace?

The objection suggested by these questions is a very valid and important one, and morphology was in an unsound state, so long as it rested upon the mere percep-

tion of the analogies which obtain between fully formed
parts. The unchecked ingenuity of speculative anatomists proved itself fully competent to spin any number
of contradictory hypotheses out of the same facts, and
endless morphological dreams threatened to supplant
scientific theory.

Happily, however, there is a criterion of morphological truth, and a sure test of all homologies. Our
lobster has not always been what we see it; it was once
an egg, a semifluid mass of yolk, not so big as a pin's
head, contained in a transparent membrane, and exhibiting not the least trace of any one of those organs,
whose multiplicity and complexity, in the adult, are so
surprising. After a time a delicate patch of cellular
membrane appeared upon one face of this yolk, and that
patch was the foundation of the whole creature, the clay
out of which it would be moulded. Gradually investing
the yolk, it became subdivided by transverse constric-
tions into segments, the forerunners of the rings of the
body. Upon the ventral surface of each of the rings
thus sketched out, a pair of bud-like prominences made
their appearance—the rudiments of the appendages of
the ring. At first, all the appendages were alike, but, as
they grew, most of them became distinguished into a
stem and two terminal divisions, to which, in the middle
part of the body, was added a third outer division; and
it was only at a later period, that by the modification, or
abortion, of certain of these primitive constituents, the
limbs acquired their perfect form.

Thus the study of development proves that the doctrine of unity of plan is not merely a fancy, that it is
not merely one way of looking at the matter, but that it

is the expression of deep-seated natural facts. The legs and jaws of the lobster may not merely be regarded as modifications of a common type,—in fact and in nature they are so,—the leg and the jaw of the young animal being, at first, indistinguishable.

These are wonderful truths, the more so because the zoologist finds them to be of universal application. The investigation of a polype, of a snail, of a fish, of a horse, or of a man, would have led us, though by a less easy path, perhaps, to exactly the same point. Unity of plan everywhere lies hidden under the mask of diversity of structure—the complex is everywhere evolved out of the simple. Every animal has at first the form of an egg, and every animal and every organic part, in reaching its adult state, passes through conditions common to other animals and other adult parts ; and this leads me to another point. I have hitherto spoken as if the lobster were alone in the world, but, as I need hardly remind you, there are myriads of other animal organisms. Of these, some, such as men, horses, birds, fishes, snails, slugs, oysters, corals, and sponges, are not in the least like the lobster. But other animals, though they may differ a good deal from the lobster, are yet either very like it, or are like something that is like it. The cray fish, the rock lobster, and the prawn, and the shrimp, for example, however different, are yet so like lobsters, that a child would group them as of the lobster kind, in contradistinction to snails and slugs ; and these last again would form a kind by themselves, in contradistinction to cows, horses, and sheep, the cattle kind.

But this spontaneous grouping into ".kinds" is the first essay of the human mind at classification, or the calling by a common name of those things that are alike, and the arranging them in such a manner as best to suggest the sum of their likenesses and unlikenesses to other things.

Those kinds which include no other subdivisions than the sexes, or various breeds, are called, in technical language, species. The English lobster is a species, our cray fish is another, our prawn is another. In other countries, however, there are lobsters, cray fish, and prawns, very like ours, and yet presenting sufficient differences to deserve distinction. Naturalists, therefore, express this resemblance and this diversity by grouping them as distinct species of the same " genus." But the lobster and the cray fish, though belonging to distinct genera, have many features in common, and hence are grouped together in an assemblage which is called a family. More distant resemblances connect the lobster with the prawn and the crab, which are expressed by putting all these into the same order. Again, more remote, but still very definite, resemblances unite the lobster with the woodlouse, the king crab, the water-flea, and the barnacle, and separate them from all other animals ; whence they collectively constitute the larger group, or class, *Crustacea.* But the *Crustacea* exhibit many peculiar features in common with insects, spiders, and centipedes, so that these are grouped into the still larger assemblage or "province" *Articulata ;* and, finally, the relations which these have to worms and other lower animals, are expressed by combining the whole vast aggregate into the sub-kingdom of *Annulosa.*

If I had worked my way from a sponge instead of a lobster, I should have found it associated, by like ties, with a great number of other animals into the sub-kingdom *Protozoa* ; if I had selected a fresh-water polype or a coral, the members of what naturalists term the sub-kingdom *Cœlenterata* would have grouped themselves around my type ; had a snail been chosen, the inhabitants of all univalve and bivalve, land and water, shells, the lamp shells, the squids, and the sea-mat would have gradually linked themselves on to it as members of the same sub-kingdom of *Mollusca ;* and finally, starting from man, I should have been compelled to admit first, the ape, the rat, the horse, the dog, into the same class ; and then the bird, the crocodile, the turtle, the frog, and the fish, into the same sub-kingdom of *Vertebrata.*

And if I had followed out all these various lines of classification fully, I should discover in the end that there was no animal, either recent or fossil, which did not at once fall into one or other of these sub-kingdoms. In other words, every animal is organized upon one or other of the five, or more, plans, whose existence renders our classification possible. And so definitely and pre-cisely marked is the structure of each animal, that, in the present state of our knowledge, there is not the least evidence to prove that a form, in the slightest degree transitional between any of the two groups *Vertebrata, Annulosa, Mollusca,* and *Cœlenterata,* either exists, or has existed, during that period of the earth's history which is recorded by the geologist. Nevertheless, you must not for a moment suppose, because no such transitional forms are known, that the members of the sub-kingdoms are disconnected from, or indepen-dent of, one another. On the contrary, in their earliest condition they are all alike, and the primordial germs of a man, a dog, a bird, a fish, a beetle, a snail, and a polype are, in no essential structural respects, distinguishable.

In this broad sense, it may with truth be said, that all living animals, and all those dead creations which geology reveals, are bound together by an all-pervading unity of organization, of the same character, though not equal in degree, to that which enables us to discern one and the same plan amidst the twenty different segments of a lobster's body. Truly it has been said, that to a clear eye the smallest fact is a window through which the Infinite may be seen.

Turning from these purely morphological considerations, let us now examine into the manner in which the attentive study of the lobster impels us into other lines of research.

Lobsters are found in all the European seas ; but on the opposite shores of the Atlantic and in the seas of the southern hemisphere they do not exist. They are, however, represented in these regions by very closely allied, but distinct forms—the *Homarus Americanus* and the *Homarus Capensis :* so that we may say that the European has one species of *Homarus ;* the American, another ; the African, another ; and thus the remarkable facts of geographical distribution begin to dawn upon us.

Again, if we examine the contents of the earth's crust, we shall find in the latter of those deposits, which have served as the great burying grounds of past ages, numberless lobster-like animals, but none so similar to our living lobster as to make zoologists sure that they belonged even to the same genus. If we go still further back in time, we discover, in the oldest rocks of all, the remains of animals, constructed on the same general plan as the lobster, and belonging to the same great group of *Crustacea;* but for the most part totally different from the lobster, and indeed from any other living form of crustacean ; and thus we gain a notion of that successive change of the animal population of the globe, in past ages, which is the most striking fact revealed by geology.

Consider, now, where our inquiries have led us. We studied our type morphologically, when we determined its anatomy and its development, and when comparing it, in these respects, with other animals, we made out its place in a system of classification. If we were to examine every animal in a similar manner, we should establish a complete body of zoological morphology.

Again, we investigated the distribution of our type in space and in time, and, if the like had been done with every animal, the sciences of geographical and geological distribution would have attained their limit.

But you will observe one remarkable circumstance, that, up to this point, the question of the life of these organisms has not come under consideration. Morphology and distribution might be studied almost as well, if animals and plants were a peculiar kind of crystals, and possessed none of those functions which distinguish living beings so remarkably. But the facts of morphology and distribution have to be accounted for, and the science, whose aim it is to account for them, is Physiology.

Let us return to our lobster once more. If we watched the creature in its native element, we should see it climbing actively the submerged rocks, among which it delights to live, by means of its strong legs; or swimming by powerful strokes of its great tail, the appendages of whose sixth joint are spread out into a broad fan-like propeller: seize it, and it will show you that its great claws are no mean weapons of offence; suspend a piece of carrion among its haunts, and it will greedily devour it, tearing and crushing the flesh by means of its multitudinous jaws.

Suppose that we had known nothing of the lobster but as an inert mass, an organic crystal, if I may use the phrase, and that we could suddenly see it exerting all these powers, what wonderful new ideas and new questions would arise in our minds! The great new question would be, "How does all this take place?" the chief new idea would be, the idea of adaptation to purpose,—the notion, that the constituents of animal bodies are not mere unconnected parts, but organs working together to an end. Let us consider the tail of the lobster again from this point of view. Morphology has taught us that it is a series of segments composed of homologous parts, which undergo various modifications—beneath and through which a common plan of formation is discernible. But if I look at the same part physiologically, I see that it is a most beautifully constructed organ of locomotion, by means of which the animal can swiftly propel itself either backwards or forwards.

But how is this remarkable propulsive machine made to perform its functions? If I were suddenly to kill one of these animals and to take out all the soft parts, I should find the shell to be perfectly inert, to have no more power of moving itself than is possessed by the machinery of a mill, when disconnected from its steam-engine or water-wheel. But if I were to open it, and take out the viscera only, leaving the white flesh, I should perceive that the lobster could bend and extend its tail as well as before. If I were to cut off the tail, I should cease to find any spontaneous motion in it; but on pinching any portion of the flesh, I should observe that it underwent a very curious change—each fibre becoming shorter and thicker. By this act of contraction, as it is termed, the parts to which the ends of the fibre are attached are, of course, approximated; and according to the relations of their points of attachment to the centres of motion of the different rings, the bending or the extension of the tail results. Close observation of the newly opened lobster would soon show that all its movements are due to the same cause—the shortening and thickening of these fleshy fibres, which are technically called muscles.

Here, then, is a capital fact. The movements of the lobster are due to muscular contractility. But why does a muscle contract at one time and not at another? Why does one whole group of muscles contract when the lobster wishes to extend his tail, and another group, when he desires to bend it? What is it originates, directs, and controls the motive power?

Experiment, the great instrument for the ascertainment of truth in physical science, answers this question for us. In the head of the lobster there lies a small mass of that peculiar tissue which is known as nervous substance. Cords of similar matter connect this brain of the lobster, directly or indirectly, with the muscles. Now, if these communicating cords are cut, the brain remaining entire, the power of exerting what we call voluntary motion in the parts below the section is destroyed; and on the other hand, if, the cords remaining entire, the brain mass be destroyed, the same voluntary mobility is equally lost. Whence the inevitable conclusion is, that the power of originating these motions resides in the brain, and is propagated along the nervous cords.

In the higher animals the phænomena which attend this transmission have been investigated, and the exertion of the peculiar energy which resides in the nerves has been found to be accompanied by a disturbance of the electrical state of their molecules.

If we could exactly estimate the signification of this disturbance; if we could obtain the value of a given exertion of nerve force by determining the quantity of electricity, or of heat, of which it is the equivalent; if we could ascertain upon what arrangement, or other condition of the molecules of matter, the manifestation of the nervous and muscular energies depends, (and doubtless science will some day or other ascertain these points,) physiologists would have attained their ultimate goal in this direction ; they would have determined the relation of the motive force of animals to the other forms of force found in nature ; and if the same process had been successfully performed for all the operations which are carried on in, and by, the animal frame, physiology would be perfect, and the facts of morphology and distribution would be deducible from the laws which physiologists had established, combined with those determining the condition of the surrounding universe.

There is not a fragment of the organism of this humble animal, whose study would not lead us into regions of thought as large as those which I have briefly opened up to you; but what I have been saying, I trust, has not only enabled you to form a conception of the scope and purport of zoology, but has given you an imperfect example of the manner in which, in my opinion, that science, or indeed any physical science, may be best taught. The great matter is, to make teaching real and practical, by fixing the attention of the student on particular facts; but at the same time it should be rendered broad and comprehensive, by constant reference to the generalizations of which all particular facts are illustrations. The lobster has served as a type of the whole animal kingdom, and its anatomy and physiology have illustrated for us some of the greatest truths of biology. The student who has once seen for himself the facts which I have described, has had their relations explained to him, and has clearly comprehended them, has, so far, a knowledge of zoology, which is real and genuine, however limited it may be, and which is worth more than all the mere reading knowledge of the science he could ever acquire. His zoological information is, so far, knowledge and not mere hearsay.

And if it were my business to fit you for the certificate in zoological science granted by this department, I should pursue a course precisely similar in principle to that which I have taken to-night. I should select a fresh-water sponge, a fresh-water polype or a *Cyanœa*, a fresh-water mussel, a lobster, a fowl, as types of the five primary divisions of the animal kingdom. I should explain their structure very fully, and show how each illustrated the great principles of zoology. Having gone very carefully and fully over this ground, I should feel that you had a safe foundation, and I should then take you in the same way, but less minutely, over similarly selected illustrative types of the classes; and then I should direct your attention to the special forms enumerated under the head of types, in this syllabus, and to the other facts there mentioned.

That would, speaking generally, be my plan. But I have undertaken to explain to you the best mode of acquiring and communicating a knowledge of zoology, and you may therefore fairly ask me for a more detailed and precise account of the manner in which I should propose to furnish you with the information I refer to.

My own impression is, that the best model for all kinds of training in physical science is that afforded by the method of teaching anatomy, in use in the medical schools. This method consists of three elements —lectures, demonstrations, and examinations.

The object of lectures is, in the first place, to awaken the attention and excite the enthusiasm of the student; and this, I am sure, may be effected to a far greater extent by the oral discourse and by the personal influence of a respected teacher than in any other way. Secondly, lectures have the double use of guiding the student to the salient points of a subject, and at the same time forcing him to attend to the whole of it, and not merely to that part which takes his fancy. And lastly, lectures afford the student the opportunity of seeking explanations of those difficulties which will, and indeed ought to, arise in the course of his studies.

But for a student to derive the utmost possible value from lectures, several precautions are needful.

I have a strong impression that the better a discourse is, as an oration, the worse it is as a lecture. The flow of the discourse carries you on without proper attention to its sense; you drop a word or a phrase, you lose the exact meaning for a moment, and while you strive to recover yourself, the speaker has passed on to something else.

The practice I have adopted of late years, in lecturing to students, is to condense the substance of the hour's discourse into a few dry propositions, which are read slowly and taken down from dictation; the reading of each being followed by a free commentary, expanding and illustrating the proposition, explaining terms, and removing any difficulties that may be attackable in that way, by diagrams made roughly, and seen to grow under the lecturer's hand. In this manner you, at any rate, insure the co-operation of the student to a certain extent. He cannot leave the lecture-room entirely empty if the taking of notes is enforced; and a student must be preternaturally dull and mechanical, if he can take notes and hear them properly explained, and yet learn nothing.

What books shall I read? is a question constantly put by the student to the teacher. My reply usually is, "None: write your notes out carefully and fully; strive to understand them thoroughly; come to me for the explanation of anything you cannot understand; and I would rather you did not distract your mind by reading." A properly composed course of lectures ought to contain fully as much matter as a student can assimilate in the time occupied by its delivery; and the teacher should always recollect that his business is to feed, and not to cram the intellect. Indeed, I believe that a student who gains from a course of lectures the simple habit of concentrating his attention upon a definitely limited series of facts, until they are thoroughly mastered, has made a step of immeasurable importance.

But, however good lectures may be, and however extensive the course of reading by which they are followed up, they are but accessories to the great instrument of scientific teaching—demonstration. If I insist unweariedly, nay fanatically, upon the importance of physical science as an educational agent, it is because the study of any branch of science, if properly conducted, appears to me to fill up a void left by all other means of education. I have the greatest respect and love for literature; nothing would grieve me more than to see literary training other than a very prominent branch of education: indeed, I wish that real literary discipline were far more attended to than it is; but I cannot shut my eyes to the fact, that there is a vast difference between men who have had a purely literary, and those who have had a sound scientific, training.

Seeking for the cause of this difference, I imagine I can find it in the fact, that, in the world of letters, learning and knowledge are one, and books are the source of both; whereas in science, as in life, learning and knowledge are distinct, and the study of things, and not of books, is the source of the latter.

All that literature has to bestow may be obtained by reading and by practical exercise in writing, and in speaking; but I do not exaggerate when I say, that none of the best gifts of science are to be won by these means. On the contrary, the great benefit which a scientific education bestows, whether as training or as knowledge, is dependent upon the extent to which the mind of the student is brought into immediate contact with facts—upon the degree to which he learns the habit of appealing directly to Nature, and of acquiring through his senses concrete images of those properties of things, which are, and always will be, but approximatively expressed in human language. Our way of looking at Nature, and of speaking about her, varies from year to year; but a fact once seen, a relation of cause and effect, once demonstratively apprehended, are possessions which neither change nor pass away, but, on the contrary, form fixed centres, about which other truths aggregate by natural affinity.

Therefore, the great business of the scientific teacher is, to imprint the fundamental, irrefragable facts of his science, not only by words upon the mind, but by sensible impressions upon the eye, and ear, and touch of the student, in so complete a manner, that every term used, or law enunciated, should afterwards call up vivid images of the particular structural, or other, facts which furnished ·the demonstration of the law, or the illustration of the term.

Now this important operation can only be achieved by constant demonstration, which may take place to a certain imperfect extent during a lecture, but which ought also to be carried on independently, and which should be addressed to each individual student, the teacher endeavouring, not so much to show a thing to the learner, as to make him see it for himself.

I am well aware that there are great practical difficulties in the way of effectual zoological demonstrations.

The dissection of animals is not altogether pleasant, and requires much time; nor is it easy to secure an adequate supply of the needful specimens. The botanist has here a great advantage; his specimens are easily obtained, are clean and wholesome, and can be dissected in a private house as well as anywhere else; and hence, I believe, the fact, that botany is so much more readily and better taught than its sister science. But, be it difficult or be it easy, if zoological science is to be properly studied, demonstration, and, consequently, dissection, must be had. Without it, no man can have a really sound knowledge of animal organization.

A good deal may be done, however, without actual dissection on the student's part, by demonstration upon specimens and preparations; and in all probability it would not be very difficult, were the demand sufficient, to organize collections of such objects, sufficient for all the purposes of elementary teaching, at a comparatively cheap rate. Even without these, much might be effected, if the zoological collections, which are open to the public, were arranged according to what has been termed the "typical principle;" that is to say, if the specimens exposed to public view were so selected, that the public could learn something from them, instead of being, as at present, merely confused by their multiplicity. For example, the grand ornithological gallery at the British Museum contains between two and three thousand species of birds, and sometimes five or six specimens of a species. They are very pretty to look at, and some of the cases are, indeed, splendid; but I will undertake to say, that no man but a professed ornithologist has ever gathered much information from the collection. Certainly, no one of the tens of thousands of the general public who have walked through that gallery ever knew more about the essential peculiarities of birds when he left the gallery, than when he entered it. But if, somewhere in that vast hall, there were a few preparations, exemplifying the leading structural peculiarities and the mode of development of a common fowl; if the types of the genera, the leading modifications in the skeleton, in the plumage at various ages, in the mode of nidification, and the like, among birds, were displayed; and if the other specimens were put away in a place where the men of science, to whom they are alone useful, could have free access to them, I can conceive that this collection might become a great instrument of scientific education.

The last implement of the teacher to which I have adverted is examination—a means of education now so thoroughly understood that I need hardly enlarge upon it. I hold that both written and oral examinations are indispensable, and, by requiring the description of specimens, they may be made to supplement demonstration.

Such is the fullest reply the time at my disposal will allow me to give to the question—how may a knowledge of zoology be best acquired and communicated?

But there is a previous question which may be moved, and which, in fact, I know many are inclined to move. It is the question, why should training masters be encouraged to acquire a knowledge of this, or any other branch of physical science? What is the use, it is said, of attempting to make physical science a branch of

primary education? It is not probable that teachers,
in pursuing such studies, will be led astray from the
acquirement of more important but less attractive
knowledge? And, even if they can learn something
of science without prejudice to their usefulness, what
is the good of their attempting to instil that knowledge
into boys whose real business is the acquisition of
reading, writing, and arithmetic?

These questions are, and will be, very commonly
asked, for they arise from that profound ignorance of
the value and true position of physical science, which
infests the minds of the most highly educated and
intelligent classes of the community. But if I did not
feel well assured that they are capable of being easily
and satisfactorily answered; that they have been answered over and over again; and that the time will
come when men of liberal education will blush to raise
such questions,—I should be ashamed of my position
here to-night. Without doubt, it is your great and very
important function to carry out elementary education;
without question, anything that should interfere with
the faithful fulfilment of that duty on your part would
be a great evil; and if I thought that your acquirement
of the elements of physical science, and your communication of those elements to your pupils, involved any
sort of interference with your proper duties, I should
be the first person to protest against your being encouraged to do anything of the kind.

But is it true that the acquisition of such a knowledge of science as is proposed, and the communication
of that knowledge, are calculated to weaken your usefulness? Or may I not rather ask, is it possible for

you to discharge your functions properly without these aids ?

What is the purpose of primary intellectual education ? I apprehend that its first object is to train the young in the use of those tools wherewith men extract knowledge from the ever-shifting succession of phenomena which pass before their eyes ; and that its second object is to inform them of the fundamental laws which have been found by experience to govern the course of things, so that they may not be turned out into the world naked, defenceless, and a prey to the events they might control.

A boy is taught to read his own and other languages, in order that he may have access to infinitely wider stores of knowledge than could ever be opened to him by oral intercourse with his fellow men ; he learns to write, that his means of communication with the rest of mankind may be indefinitely enlarged, and that he may record and store up the knowledge he acquires. He is taught elementary mathematics, that he may understand all those relations of number and form, upon which the transactions of men, associated in complicated societies, are built, and that he may have some practice in deductive reasoning.

All these operations of reading, writing, and ciphering, are intellectual tools, whose use should, before all things, be learned, and learned thoroughly ; so that the youth may be enabled to make his life that which it ought to be, a continual progress in learning and in wisdom.

But, in addition, primary education endeavours to. fit a boy out with a certain equipment of positive knowledge. He is taught the great laws of morality ; the.

religion of his sect; so much history and geography as will tell him where the great countries of the world are, what they are, and how they have become what they are.

Without doubt all these are most fitting and excellent things to teach a boy; I should be very sorry to omit any of them from any scheme of primary intellectual education. The system is excellent, so far as it goes.

But if I regard it closely, a curious reflection arises. I suppose that, fifteen hundred years ago, the child of any well-to-do Roman citizen was taught just these same things; reading and writing in his own, and, perhaps, the Greek tongue; the elements of mathematics; and the religion, morality, history, and geography current in his time. Furthermore, I do not think I err in affirming, that, if such a Christian Roman boy, who had finished his education, could be transplanted into one of our public schools, and pass through its course of instruction, he would not meet with a single unfamiliar line of thought; amidst all the new facts he would have to learn, not one would suggest a different mode of regarding the universe from that current in his own time.

And yet surely there is some great difference between the civilization of the fourth century and that of the nineteenth, and still more between the intellectual habits and tone of thought of that day and this?

And what has made this difference? I answer fearlessly,—The prodigious development of physical science within the last two centuries.

Modern civilization rests upon physical science; take

K

away her gifts to our own country, and our position among the leading nations of the world is gone to-morrow; for it is physical science only, that makes intelligence and moral energy stronger than brute force.

The whole of modern thought is steeped in science; it has made its way into the works of our best poets, and even the mere man of letters, who affects to ignore and despise science, is unconsciously impregnated with her spirit, and indebted for his best products to her methods. I believe that the greatest intellectual revolution mankind has yet seen is now slowly taking place by her agency. She is teaching the world that the ultimate court of appeal is observation and experiment, and not authority; she is teaching it to estimate the value of evidence; she is creating a firm and living faith in the existence of immutable moral and physical laws, perfect obedience to which is the highest possible aim of an intelligent being.

But of all this your old stereotyped system of education takes no note. Physical science, its methods, its problems, and its difficulties, will meet the poorest boy at every turn, and yet we educate him in such a manner that he shall enter the world as ignorant of the existence of the methods and facts of science as the day he was born. The modern world is full of artillery; and we turn out our children to do battle in it, equipped with the shield and sword of an ancient gladiator.

Posterity will cry shame on us if we do not remedy this deplorable state of things. Nay, if we live twenty years longer, our own consciences will cry shame on us.

It is my firm conviction that the only way to remedy it is, to make the elements of physical science an integral part of primary education. I have endeavoured to show you how that may be done for that branch of science which it is my business to pursue ; and I can but add, that I should look upon the day when every school-master throughout this land was a centre of genuine, however rudimentary, scientific knowledge, as an epoch in the history of the country.

But let me entreat you to remember my last words. Addressing myself to you, as teachers, I would say, mere book learning in physical science is a sham and a delusion—what you teach, unless you wish to be impostors, that you must first know ; and real knowledge in science means personal acquaintance with the facts, be they few or many.[1]

---

[1] It has been suggested to me that these words may be taken to imply a discouragement on my part of any sort of scientific instruction which does not give an acquaintance with the facts at first hand. But this is not my meaning. The ideal of scientific teaching is, no doubt, a system by which the scholar sees every fact for himself, and the teacher supplies only the explanations. Circumstances, however, do not often allow of the attainment of that ideal, and we must put up with the next best system — one in which the scholar takes a good deal on trust from a teacher, who, knowing the facts by his own knowledge, can describe them with so much vividness as to enable his audience to form competent ideas concerning them. The system which I repudiate is that which allows teachers who have not come into direct contact with the leading facts of a science to pass their second-hand information on. The scientific virus, like vaccine lymph, if passed through too long a succession of organisms, will lose all its effect in protecting the young against the intellectual epidemics to which they are exposed.

VII.

ON THE PHYSICAL BASIS OF LIFE.[1]

IN order to make the title of this discourse generally intelligible, I have translated the term "Protoplasm," which is the scientific name of the substance of which I am about to speak, by the words "the physical basis of life." I suppose that, to many, the idea that there is such a thing as a physical basis, or matter, of life may be novel—so widely spread is the conception of life as a something which works through matter, but is independent of it; and even those who are aware that matter and life are inseparably connected, may not be prepared for the conclusion plainly suggested by the phrase, "*the physical basis or matter of life*," that there is some one

---

[1] The substance of this paper was contained in a discourse which was delivered in Edinburgh on the evening of Sunday, the 8th of November, 1868—being the first of a series of Sunday evening addresses upon non-theological topics, instituted by the Rev. J. Cranbrook. Some phrases, which could possess only a transitory and local interest, have been omitted; instead of the newspaper report of the Archbishop of York's address, his Grace's subsequently-published pamphlet "On the Limits of Philosophical Inquiry" is quoted; and I have, here and there, endeavoured to express my meaning more fully and clearly than I seem to have done in speaking—if I may judge by sundry criticisms upon what I am supposed to have said, which have appeared. But in substance, and, so far as my recollection serves, in form, what is here written corresponds with what was there said.

kind of matter which is common to all living beings, and that their endless diversities are bound together by a physical, as well as an ideal, unity. In fact, when first apprehended, such a doctrine as this appears almost shocking to common sense.

What, truly, can seem to be more obviously different from one another in faculty, in form, and in substance, than the various kinds of living beings? What community of faculty can there be between the brightly-coloured lichen, which so nearly resembles a mere mineral incrustation of the bare rock on which it grows, and the painter, to whom it is instinct with beauty, or the botanist, whom it feeds with knowledge?

Again, think of the microscopic fungus—a mere infinitesimal ovoid particle, which finds space and duration enough to multiply into countless millions in the body of a living fly; and then of the wealth of foliage, the luxuriance of flower and fruit, which lies between this bald sketch of a plant and the giant pine of California, towering to the dimensions of a cathedral spire, or the Indian fig, which covers acres with its profound shadow, and endures while nations and empires come and go around its vast circumference? Or, turning to the other half of the world of life, picture to yourselves the great Finner whale, hugest of beasts that live, or have lived, disporting his eighty or ninety feet of bone, muscle, and blubber, with easy roll, among waves in which the stoutest ship that ever left dockyard would founder hopelessly; and contrast him with the invisible animalcules—mere gelatinous specks, multitudes of which could, in fact, dance upon the point of a needle with the same ease as the angels of the Schoolmen could, in imagination.

With these images before your minds, you may well ask, what community of form, or structure, is there between the animalcule and the whale ; or between the fungus and the fig-tree ?   And, *à fortiori*, between all four ?

Finally, if we regard substance, or material composition, what hidden bond can connect the flower which a girl wears in her hair and the blood which courses through her youthful veins; or, what is there in common between the dense and resisting mass of the oak, or the strong fabric of the tortoise, and those broad disks of glassy jelly which may be seen pulsating through the waters of a calm sea, but which drain away to mere films in the hand which raises them out of their element ?

Such objections as these must, I think, arise in the mind of every one who ponders, for the first time, upon the conception of a single physical basis of life underlying all the diversities of vital existence ; but I propose to demonstrate to you that, notwithstanding these apparent difficulties, a threefold unity—namely, a unity of power, or faculty, a unity of form, and a unity of substantial composition—does pervade the whole living world.

No very abstruse argumentation is needed, in the first place, to prove that the powers, or faculties, of all kinds of living matter, diverse as they may be in degree, are substantially similar in kind.

Goethe has condensed a survey of all the powers of mankind into the well-known epigram :—

" Warum treibt sich das Volk so und schreit ? Es will sich ernähren
Kinder zeugen, und die nähren so gut es vermag.
*   *   *   *   *
Weiter bringt es kein Mensch, stell' er sich wie er auch will."

In physiological language this means, that all the multifarious and complicated activities of man are comprehensible under three categories. Either they are immediately directed towards the maintenance and development of the body, or they effect transitory changes in the relative positions of parts of the body, or they tend towards the continuance of the species. Even those manifestations of intellect, of feeling, and of will, which we rightly name the higher faculties, are not excluded from this classification, inasmuch as to every one but the subject of them, they are known only as transitory changes in the relative positions of parts of the body. Speech, gesture, and every other form of human action are, in the long run, resolvable into muscular contraction, and muscular contraction is but a transitory change in the relative positions of the parts of a muscle. But the scheme which is large enough to embrace the activities of the highest form of life, covers all those of the lower creatures. The lowest plant, or animalcule, feeds, grows, and reproduces its kind. In addition, all animals manifest those transitory changes of form which we class under irritability and contractility; and, it is more than probable, that when the vegetable world is thoroughly explored, we shall find all plants in possession of the same powers, at one time or other of their existence.

I am not now alluding to such phænomena, at once rare and conspicuous, as those exhibited by the leaflets of the sensitive plant, or the stamens of the barberry, but to much more widely-spread, and, at the same time, more subtle and hidden, manifestations of vegetable contractility. You are doubtless aware that the common nettle owes its stinging property to the innumerable stiff and needle-like, though exquisitely delicate, hairs which cover its surface. Each stinging-needle tapers from a broad base to a slender summit, which, though rounded at the end, is of such microscopic fineness that it readily penetrates, and breaks off in, the skin. The whole hair consists of a very delicate outer case of wood, closely applied to the inner surface of which is a layer of semi-fluid matter, full of innumerable granules of extreme minuteness. This semi-fluid lining is protoplasm, which thus constitutes a kind of bag, full of a limpid liquid, and roughly corresponding in form with the interior of the hair which it fills. When viewed with a sufficiently high magnifying power, the protoplasmic layer of the nettle hair is seen to be in a condition of unceasing activity. Local contractions of the whole thickness of its substance pass slowly and gradually from point to point, and give rise to the appearance of progressive waves, just as the bending of successive stalks of corn by a breeze produces the apparent billows of a corn-field.

But, in addition to these movements, and independently of them, the granules are driven, in relatively rapid streams, through channels in the protoplasm which seem to have a considerable amount of persistence. Most commonly, the currents in adjacent parts of the proto-plasm take similar directions ; and, thus, there is a general stream up one side of the hair and down the other. But this does not prevent the existence of partial currents which take different routes ; and, sometimes, trains of granules may be seen coursing swiftly in opposite directions, within a twenty-thousandth of an inch of one another; while, occasionally, opposite streams come into direct collision, and, after a longer or shorter struggle, one predominates. The cause of these currents seems to lie in contractions of the protoplasm which bounds the channels in which they flow, but which are so minute that the best microscopes show only their effects, and not themselves.

The spectacle afforded by the wonderful energies prisoned within the compass of the microscopic hair of a plant, which we commonly regard as a merely passive organism, is not easily forgotten by one who has watched its display, continued hour after hour, without pause or sign of weakening. The possible complexity of many other organic forms, seemingly as simple as the protoplasm of the nettle, dawns upon one; and the comparison of such a protoplasm to a body with an internal circulation, which has been put forward by an eminent physiologist, loses much of its startling character. Currents similar to those of the hairs of the nettle have been observed in a great multitude of very different plants, and weighty authorities have suggested that they probably occur, in more or less perfection, in all young vegetable cells. If such be the case, the wonderful noonday silence of a tropical forest is, after all, due only to the dulness of our hearing; and could our ears catch the murmur of these tiny Maelstroms, as they whirl in the innumerable myriads of living cells which constitute each tree, we should be stunned, as with the roar of a great city.

Among the lower plants, it is the rule rather than the exception, that contractility should be still more openly manifested at some periods of their existence. The protoplasm of *Algæ* and *Fungi* becomes, under many circumstances, partially, or completely, freed from its woody case, and exhibits movements of its whole mass, or is propelled by the contractility of one, or more, hair-like prolongations of its body, which are called vibratile cilia. And, so far as the conditions of the manifestation of the phænomena of contractility have yet been studied, they are the same for the plant as for the animal. Heat and electric shocks influence both, and in the same way, though it may be in different degrees. It is by no means my intention to suggest that there is no difference in faculty between the lowest plant and the highest, or between plants and animals. But the difference between the powers of the lowest plant, or animal, and those of the highest, is one of degree, not of kind, and depends, as Milne-Edwards long ago so well pointed out, upon the extent to which the principle of the division of labour is carried out in the living economy. In the lowest organism all parts are competent to perform all functions, and one and the same portion of proto-plasm may successively take on the function of feeding, moving, or reproducing apparatus. In the highest, on the contrary, a great number of parts combine to per-form each function, each part doing its allotted share of the work with great accuracy and efficiency, but being useless for any other purpose.

On the other hand, notwithstanding all the funda-mental resemblances which exist between the powers of the protoplasm in plants and in animals, they present a striking difference (to which I shall advert more at length presently), in the fact that plants can manufacture fresh protoplasm out of mineral compounds, whereas animals are obliged to procure it ready made, and hence, in the long run, depend upon plants. Upon what condition this difference in the powers of the two great divisions of the world of life depends, nothing is at present known.

With such qualification as arises out of the last-mentioned fact, it may be truly said that the acts of all living things are fundamentally one. Is any such unity predicable of their forms? Let us seek in easily verified facts for a reply to this question. If a drop of blood be drawn by pricking one's finger, and viewed with proper precautions and under a sufficiently high microscopic power, there will be seen, among the innumerable multitude of little, circular, discoidal bodies, or corpuscles, which float in it and give it its colour, a comparatively small number of colourless corpuscles, of somewhat larger size and very irregular shape. If the drop of blood be kept at the temperature of the body, these colourless corpuscles will be seen to exhibit a marvellous activity, changing their forms with great rapidity, drawing in and thrusting out prolongations of their substance, and creeping about as if they were independent organisms.

The substance which is thus active is a mass of protoplasm, and its activity differs in detail, rather than in principle, from that of the protoplasm of the nettle. Under sundry circumstances the corpuscle dies and becomes distended into a round mass, in the midst of which is seen a smaller spherical body, which existed, but was more or less hidden, in the living corpuscle, and is called its *nucleus*. Corpuscles of essentially similar structure are to be found in the skin, in the lining of the mouth, and scattered through the whole framework of the body. Nay, more; in the earliest condition of the human organism, in that state in which it has but just become distinguishable from the egg in which it arises, it is nothing but an aggregation of such corpuscles, and every organ of the body was, once, no more than such an aggregation.

Thus a nucleated mass of protoplasm turns out to be what may be termed the structural unit of the human body. As a matter of fact, the body, in its earliest state, is a mere multiple of such units ; and, in its perfect condition, it is a multiple of such units, variously modified.

But does the formula which expresses the essential structural character of the highest animal cover all the rest, as the statement of its powers and faculties covered that of all others ? Very nearly. Beast and fowl, reptile and fish, mollusk, worm, and polype, are all composed of structural units of the same character, namely, masses of protoplasm with a nucleus. There are sundry very low animals, each of which, structurally, is a mere colourless blood-corpuscle, leading an independent life. But, at the very bottom of the animal scale, even this simplicity becomes simplified, and all the phænomena of life are manifested by a particle of protoplasm without a nucleus. Nor are such organisms insignificant by reason of their want of complexity. It is a fair question whether the protoplasm of those simplest forms of life, which people an immense extent of the bottom of the sea, would not outweigh that of all the higher living beings which inhabit the land put together. And in ancient times, no less than at the present day, such living beings as these have been the greatest of rock builders.

What has been said of the animal world is no less true

of plants. Imbedded in the protoplasm at the broad, or attached, end of the nettle hair, there lies a spheroidal nucleus. Careful examination further proves that the whole substance of the nettle is made up of a repetition of such masses of nucleated protoplasm, each contained in a wooden case, which is modified in form, sometimes into a woody fibre, sometimes into a duct or spiral vessel, sometimes into a pollen grain, or an ovule. Traced back to its earliest state, the nettle arises as the man does, in a particle of nucleated protoplasm. And in the lowest plants, as in the lowest animals, a single mass of such protoplasm may constitute the whole plant, or the protoplasm may exist without a nucleus.

Under these circumstances it may well be asked, how is one mass of non-nucleated protoplasm to be distinguished from another? why call one "plant" and the other "animal"?

The only reply is that, so far as form is concerned, plants and animals are not separable, and that, in many cases, it is a mere matter of convention whether we call a given organism an animal or a plant. There is a living body called *Æthalium septicum*, which appears upon decaying vegetable substances, and, in one of its forms, is common upon the surfaces of tan-pits. In this condition it is, to all intents and purposes, a fungus, and formerly was always regarded as such; but the remarkable investigations of De Bary have shown that, in another condition, the *Æthalium* is an actively locomotive creature, and takes in solid matters, upon which, apparently, it feeds, thus exhibiting the most characteristic feature of animality. Is this a plant; or is it an animal? Is it both; or is it neither? Some decide in favour of the last supposition, and establish an intermediate kingdom, a sort of biological No Man's Land for all these questionable forms. But, as it is admittedly impossible to draw any distinct boundary line between this no man's land and the vegetable world on the one hand, or the animal, on the other, it appears to me that this proceeding merely doubles the difficulty which, before, was single.

Protoplasm, simple or nucleated, is the formal basis of all life. It is the clay of the potter : which, bake it and paint it as he will, remains clay, separated by artifice, and not by nature, from the commonest brick or sun-dried clod.

Thus it becomes clear that all living powers are cognate, and that all living forms are fundamentally of one character. The researches of the chemist have revealed a no less striking uniformity of material composition in living matter.

In perfect strictness, it is true that chemical investigation can tell us little or nothing, directly, of the composition of living matter, inasmuch as such matter must needs die in the act of analysis,—and upon this very obvious ground, objections, which I confess seem to me to be somewhat frivolous, have been raised to the drawing of any conclusions whatever respecting the composition of actually living matter, from that of the dead matter of life, which alone is accessible to us. But objectors of this class do not seem to reflect that it is also, in strictness, true that we know nothing about the composition of any body whatever, as it is. The statement that a crystal of calc-spar consists of carbonate of lime, is quite true, if we only mean that, by appropriate processes, it may be resolved into carbonic acid and quicklime. If you pass the same carbonic acid over the very quicklime thus obtained, you will obtain carbonate of lime again; but it will not be calc-spar, nor anything like it. Can it, therefore, be said that chemical analysis teaches nothing about the chemical composition of calc-spar? Such a statement would be absurd; but it is hardly more so than the talk one occasionally hears about the uselessness of applying the results of chemical analysis to the living bodies which have yielded them.

One fact, at any rate, is out of reach of such refinements, and this is, that all the forms of protoplasm which have yet been examined contain the four elements, carbon, hydrogen, oxygen, and nitrogen, in very complex union, and that they behave similarly towards several reagents. To this complex combination, the nature of which has never been determined with exactness, the name of Protein has been applied. And if we use this term with such caution as may properly arise out of our comparative ignorance of the things for which it stands, it may be truly said, that all protoplasm is proteinaceous; or, as the white, or albumen, of an egg is one of the commonest examples of a nearly pure proteine matter, we may say that all living matter is more or less albuminoid.

Perhaps it would not yet be safe to say that all forms of protoplasm are affected by the direct action of electric shocks; and yet the number of cases in which the contraction of protoplasm is shown to be effected by this agency increases every day.

Nor can it be affirmed with perfect confidence, that all forms of protoplasm are liable to undergo that peculiar coagulation at a temperature of 40°—50° centigrade, which has been called "heat-stiffening," though Kühne's beautiful researches have proved this occurrence to take place in so many and such diverse living beings, that it is hardly rash to expect that the law holds good for all.

Enough has, perhaps, been said to prove the existence of a general uniformity in the character of the proto-plasm, or physical basis, of life, in whatever group of living beings it may be studied. But it will be under-stood that this general uniformity by no means excludes any amount of special modifications of the fundamental substance. The mineral, carbonate of lime, assumes an immense diversity of characters, though no one doubts that, under all these Protean changes, it is one and the same thing.

And now, what is the ultimate fate, and what the origin, of the matter of life ?

Is it, as some of the older naturalists supposed, diffused throughout the universe in molecules, which are indestructible and unchangeable in themselves ; but, in endless transmigration, unite in innumerable permu-tations, into the diversified forms of life we know ? Or, is the matter of life composed of ordinary matter, differing from it only in the manner in which its atoms are aggregated ? Is it built up of ordinary matter, and again resolved into ordinary matter when its work is done ?

Modern science does not hesitate a moment between these alternatives. Physiology writes over the portals of life—

"Debemur morti nos nostraque,"

with a profounder meaning than the Roman poet attached to that melancholy line. Under whatever disguise it takes refuge, whether fungus or oak, worm or man, the living protoplasm not only ultimately dies and is resolved into its mineral and lifeless constituents, but is always dying, and, strange as the paradox may sound, could not live unless it died.

In the wonderful story of the " Peau de Chagrin," the hero becomes possessed of a magical wild ass' skin, which yields him the means of gratifying all his wishes. But its surface represents the duration of the proprietor's life ; and for every satisfied desire the skin shrinks in proportion to the intensity of fruition, until at length life and the last handbreadth of the *peau de chagrin* disappear with the gratification of a last wish.

Balzac's studies had led him over a wide range of thought and speculation, and his shadowing forth of physiological truth in this strange story may have been intentional. At any rate, the matter of life is a veritable *peau de chagrin*, and for every vital act it is somewhat the smaller. All work implies waste, and the work of life results, directly or indirectly, in the waste of protoplasm.

Every word uttered by a speaker costs him some physical loss ; and, in the strictest sense, he burns that others may have light—so much eloquence, so much of his body resolved into carbonic acid, water, and urea. It is clear that this process of expenditure cannot go on for ever. But happily, the protoplasmic *peau de chagrin* differs from Balzac's in its capacity of being repaired, and brought back to its full size, after every exertion.

For example, this present lecture, whatever its intellectual worth to you, has a certain physical value to me, which is, conceivably, expressible by the number of grains of protoplasm and other bodily substance wasted in maintaining my vital processes during its delivery. My *peau de chagrin* will be distinctly smaller at the end of the discourse than it was at the beginning. By and by, I shall probably have recourse to the substance commonly called mutton, for the purpose of stretching it back to its original size. Now this mutton was once the living protoplasm, more or less modified, of another animal—a sheep. As I shall eat it, it is the same matter altered, not only by death, but by exposure to sundry artificial operations in the process of cooking.

But these changes, whatever be their extent, have not rendered it incompetent to resume its old functions as matter of life. A singular inward laboratory, which I possess, will dissolve a certain portion of the modified protoplasm; the solution so formed will pass into my veins; and the subtle influences to which it will then be subjected will convert the dead protoplasm into living protoplasm, and transubstantiate sheep into man.

Nor is this all. If digestion were a thing to be trifled with, I might sup upon lobster, and the matter of life of the crustacean would undergo the same wonderful metamorphosis into humanity. And were I to return to my own place by sea, and undergo shipwreck, the crustacea might, and probably would, return the compliment, and demonstrate our common nature by turning my protoplasm into living lobster. Or, if nothing better were to be had, I might supply my wants with mere bread, and I should find the protoplasm of the wheat-plant to be convertible into man, with no more trouble than that of the sheep, and with far less, I fancy, than that of the lobster.

Hence it appears to be a matter of no great moment what animal, or what plant, I lay under contribution for protoplasm, and the fact speaks volumes for the general identity of that substance in all living beings. I share this catholicity of assimilation with other animals, all of which, so far as we know, could thrive equally well on the protoplasm of any of their fellows, or of any plant; but here the assimilative powers of the animal world cease. A solution of smelling-salts in water, with an infinitesimal proportion of some other saline matters, contains all the elementary bodies which enter into the composition of protoplasm; but, as I need hardly say, a hogshead of that fluid would not keep a hungry man from starving, nor would it save any animal whatever from a like fate. An animal cannot make protoplasm, but must take it ready-made from some other animal, or some plant—the animal's highest feat of constructive chemistry being to convert dead protoplasm into that living matter of life which is appropriate to itself.

Therefore, in seeking for the origin of protoplasm, we must eventually turn to the vegetable world. The fluid containing carbonic acid, water, and ammonia, which offers such a Barmecide feast to the animal, is a table richly spread to multitudes of plants; and, with a due supply of only such materials, many a plant will not only maintain itself in vigour, but grow and multiply, until it has increased a million-fold, or a million million-fold, the quantity of protoplasm which it originally possessed; in this way building up the matter of life, to an indefinite extent, from the common matter of the universe.

Thus, the animal can only raise the complex substance of dead protoplasm to the higher power, as one may say, of living protoplasm ; while the plant can raise the less complex substances—carbonic acid, water, and ammonia—to the .same stage of living protoplasm, if not to the same level. But the plant also has its limitations. Some of the fungi, for example, appear to need higher compounds to start with ; and no known plant can live upon the uncompounded elements of protoplasm. A plant supplied with pure carbon, hydrogen, oxygen, and nitrogen, phosphorus, sulphur, and the like, would as infallibly die as the animal in his bath of smelling-salts, though it would be surrounded by all the constituents of protoplasm. Nor, indeed, need the process of simplification of vegetable food be carried so far as this, in order to arrive at the limit of the plant's thaumaturgy. Let water, carbonic acid, and all the other needful constituents be supplied with ammonia, and an ordinary plant will still be unable to manufacture protoplasm.

Thus the matter of life, so far as we know it (and we have no right to speculate on any other), breaks up, in consequence of that continual death which is the condition of its manifesting vitality, into carbonic acid, water, and ammonia, which certainly possess no properties but those of ordinary matter. And out of these same forms of ordinary matter, and from none which are simpler, the vegetable world builds up all the protoplasm which keeps the animal world a going. Plants are the accumulators of the power which animals distribute and disperse.

But it will be observed, that the existence of the matter of life depends on the pre-existence of certain compounds ; namely, carbonic acid, water, and ammonia.

Withdraw any one of these three from the world and all vital phænomena come to an end. They are related to the protoplasm of the plant, as the protoplasm of the plant is to that of the animal. Carbon, hydrogen, oxygen and nitrogen are all lifeless bodies. Of these, carbon and oxygen unite, in certain proportions and under certain conditions, to give rise to carbonic acid; hydrogen and oxygen produce water; nitrogen and hydrogen give rise to ammonia. These new compounds like the elementary bodies of which they are composed, are lifeless. But when they are brought together, under certain conditions they give rise to the still more complex body, protoplasm, and this protoplasm exhibits the phænomena of life.

I see no break in this series of steps in molecular complication, and I am unable to understand why the language which is applicable to any one term of the series may not be used to any of the others. We think fit to call different kinds of matter carbon, oxygen, hydrogen, and nitrogen, and to speak of the various powers and activities of these substances as the properties of the matter of which they are composed.

When hydrogen and oxygen are mixed in a certain proportion, and an electric spark is passed through them, they disappear, and a quantity of water, equal in weight to the sum of their weights, appears in their place. There is not the slightest parity between the passive and active powers of the water and those of the oxygen and hydrogen which have given rise to it. At 32° Fahrenheit, and far below that temperature, oxygen and hydrogen are elastic gaseous bodies, whose particles tend to rush away from one another with great force. Water, at the same temperature, is a strong though brittle solid, whose particles tend to cohere into definite geometrical shapes, and sometimes build up frosty imitations of the most complex forms of vegetable foliage.

Nevertheless we call these, and many other strange phænomena, the properties of the water, and we do not hesitate to believe that, in some way or another, they result from the properties of the component elements of the water. We do not assume that a something called " aquosity " entered into and took possession of the oxide of hydrogen as soon as it was formed, and then guided the aqueous particles to their places in the facets of the crystal, or amongst the leaflets of the hoar-frost. On the contrary, we live in the hope and in the faith that, by the advance of molecular physics, we shall by and by be able to see our way as clearly from the constituents of water to the properties of water, as we are now able to deduce the operations of a watch from the form of its parts and the manner in which they are put together.

Is the case in any way changed when carbonic acid, water, and ammonia disappear, and in their place, under the influence of pre-existing living protoplasm, an equivalent weight of the matter of life makes its appearance ?

It is true that there is no sort of parity between the properties of the components and the properties of the resultant, but neither was there in the case of the water. It is also true that what I have spoken of as the influence of pre-existing living matter is something quite unintelligible ; but does anybody quite comprehend the *modus operandi* of an electric spark, which traverses a mixture of oxygen and hydrogen ?

What justification is there, then, for the assumption of the existence in the living matter of a something which has no representative, or correlative, in the not living matter which gave rise to it? What better philosophical status has "vitality" than "aquosity"? And why should "vitality" hope for a better fate than the other "itys" which have disappeared since Martinus Scriblerus accounted for the operation of the meat-jack by its inherent "meat roasting quality," and scorned the "materialism" of those who explained the turning of the spit by a certain mechanism worked by the draught of the chimney?

If scientific language is to possess a definite and constant signification whenever it is employed, it seems to me that we are logically bound to apply to the protoplasm, or physical basis of life, the same conceptions as those which are held to be legitimate elsewhere. If the phænomena exhibited by water are its properties, so are those presented by protoplasm, living or dead, its properties.

If the properties of water may be properly said to result from the nature and disposition of its component molecules, I can find no intelligible ground for refusing to say that the properties of protoplasm result from the nature and disposition of its molecules.

But I bid you beware that, in accepting these conclusions, you are placing your feet on the first rung of a ladder which, in most people's estimation, is the reverse of Jacob's, and leads to the antipodes of heaven. It may seem a small thing to admit that the dull vital actions of a fungus, or a foraminifer, are the properties of their protoplasm, and are the direct results of the nature of the matter of which they are composed. But if, as I have endeavoured to prove to you, their protoplasm is essentially identical with, and most readily converted into, that of any animal, I can discover no logical halting-place between the admission that such is the case, and the further concession that all vital action may, with equal propriety, be said to be the result of the molecular forces of the protoplasm which displays it. And if so, it must be true, in the same sense and to the same extent, that the thoughts to which I am now giving utterance, and your thoughts regarding them, are the expression of molecular changes in that matter of life which is the source of our other vital phænomena.

Past experience leads me to be tolerably certain that, when the propositions I have just placed before you are accessible to public comment and criticism, they will be condemned by many zealous persons, and perhaps by some few of the wise and thoughtful. I should not wonder if "gross and brutal materialism" were the mildest phrase applied to them in certain quarters. And, most undoubtedly, the terms of the propositions are distinctly materialistic. Nevertheless two things are certain: the one, that I hold the statements to be substantially true; the other, that I, individually, am no materialist, but, on the contrary, believe materialism to involve grave philosophical error.

This union of materialistic terminology with the repudiation of materialistic philosophy, I share with some of the most thoughtful men with whom I am acquainted. And, when I first undertook to deliver the present discourse, it appeared to me to be a fitting opportunity to explain how such a union is not only consistent with, but necessitated by, sound logic. I purposed to lead you through the territory of vital phænomena to the materialistic slough in which you find yourselves now plunged, and then to point out to you the sole path by which, in my judgment, extrication is possible.

An occurrence of which I was unaware until my arrival here last night, renders this line of argument singularly opportune. I found in your papers the eloquent address " On the Limits of Philosophical Inquiry," which a distinguished prelate of the English Church delivered before the members of the Philosophical Institution on the previous day. My argument, also, turns upon this very point of the limits of philosophical inquiry ; and I cannot bring out my own views better than by contrasting them with those so plainly, and, in the main, fairly, stated by the Archbishop of York.

But I may be permitted to make a preliminary comment upon an occurrence that greatly astonished me. Applying the name of " the New Philosophy " to that estimate of the limits of philosophical inquiry which I, in common with many other men of science, hold to be just, the Archbishop opens his address by identifying this " New Philosophy " with the Positive Philosophy of M. Comte (of whom he speaks as its " founder ") ; and then proceeds to attack that philosopher and his doctrines vigorously.

Now, so far as I am concerned, the most reverend prelate might dialectically hew M. Comte in pieces, as a modern Agag, and I should not attempt to stay his hand. In so far as my study of what specially characterises the Positive Philosophy has led me, I find therein little or nothing of any scientific value, and a great deal which is as thoroughly antagonistic to the very essence of science as anything in ultramontane Catholicism.   In fact, M. Comte's philosophy in practice might be compendiously described as Catholicism *minus* Christianity.

But what has Comtism to do with the "New Philosophy," as the Archbishop defines it in the following passage ?

" Let me briefly remind you of the leading principles of this new philosophy.

" All knowledge is experience of facts acquired by the senses.   The traditions of older philosophies have obscured our experience by mixing with it much that the senses cannot observe, and until these additions are discarded our knowledge is impure.   Thus metaphysics tell us that one fact which we observe is a cause, and another is the effect of that cause ; but upon a rigid analysis, we find that our senses observe nothing of cause or effect : they observe, first, that one fact succeeds another, and, after some opportunity, that this fact has never failed to follow—that for cause and effect we should substitute invariable succession.   An older philosophy teaches us to define an object by distinguishing its essential from its accidental qualities : but experience knows nothing of essential and accidental ; she sees only that certain marks attach to an object, and, after many observations, that some of them attach invariably, whilst others may at times be absent. . . . . . As all knowledge is relative, the notion of anything being necessary must be banished with other traditions." [1]

There is much here that expresses the spirit of the " New Philosophy," if by that term be meant the spirit of modern science ; but I cannot but marvel that the assembled wisdom and learning of Edinburgh should have uttered no sign of dissent, when Comte was declared to be the founder of these doctrines.   No one will accuse Scotchmen of habitually forgetting their great countrymen ; but it was enough to make David

[1] "The Limits of Philosophical Inquiry," pp. 4 and 5.

Hume turn in his grave, that here, almost within ear-shot of his house, an instructed audience should have listened, without a murmur, while his most characteristic doctrines were attributed to a French writer of fifty years later date, in whose dreary and verbose pages we miss alike the vigour of thought and the exquisite clearness of style of the man whom I make bold to term the most acute thinker of the eighteenth century—even though that century produced Kant.

But I did not come to Scotland to vindicate the honour of one of the greatest men she has ever produced. My business is to point out to you that the only way of escape out of the crass materialism in which we just now landed, is the adoption and strict working-out of the very principles which the Archbishop holds up to reprobation.

Let us suppose that knowledge is absolute, and not relative, and therefore, that our conception of matter represents that which it really is. Let us suppose, further, that we do know more of cause and effect than a certain definite order of succession among facts, and that we have a knowledge of the necessity of that succession—and hence, of necessary laws—and I, for my part, do not see what escape there is from utter materialism and necessarianism. For it is obvious that our knowledge of what we call the material world, is, to begin with, at least as certain and definite as that of the spiritual world, and that our acquaintance with law is of as old a date as our knowledge of spontaneity. Further, I take it to be demonstrable that it is utterly impossible to prove that anything whatever may not be the effect of a material and necessary cause, and that human logic is equally incompetent to prove that any act is really spontaneous. A really spontaneous act is one which, by the assumption, has no cause; and the attempt to prove such a negative as this is, on the face of the matter, absurd. And while it is thus a philosophical impossibility to demonstrate that any given phænomenon is not the effect of a material cause, any one who is acquainted with the history of science will admit, that its progress has, in all ages, meant, and now, more than ever, means, the extension of the province of what we call matter and causation, and the concomitant gradual banishment from all regions of human thought of what we call spirit and spontaneity.

I have endeavoured, in the first part of this discourse, to give you a conception of the direction towards which modern physiology is tending ; and I ask you, what is the difference between the conception of life as the product of a certain disposition of material molecules, and the old notion of an Archæus governing and directing blind matter within each living body, except this—that here, as elsewhere, matter and law have devoured spirit and spontaneity ? And as surely as every future grows out of past and present, so will the physiology of the future gradually extend the realm of matter and law until it is co-extensive with knowledge, with feeling, and with action.

The consciousness of this great truth weighs like a nightmare, I believe, upon many of the best minds of these days. They watch what they conceive to be the progress of materialism, in such fear and powerless anger as a savage feels, when, during an eclipse, the great shadow creeps over the face of the sun. The advancing tide of matter threatens to drown their souls; the tightening grasp of law impedes their freedom; they are alarmed lest man's moral nature be debased by the increase of his wisdom.

If the " New Philosophy " be worthy of the reprobation with which it is visited, I confess their fears seem to me, to be well founded. While, on the contrary, could David Hume be consulted, I think he would smile at their perplexities, and chide them for doing even as the heathen, and falling down in terror before the hideous idols their own hands have raised.

For, after all, what do we know of this terrible " matter," except as a name for the unknown and hypothetical cause of states of our own consciousness? And what do we know of that " spirit " over whose threatened extinction by matter a great lamentation is arising, like that which was heard at the death of Pan, except that it is also a name for an unknown and hypothetical cause, or condition, of states of consciousness? In other words, matter and spirit are but names for the imaginary substrata of groups of natural phænomena.

And what is the dire necessity and " iron " law under which men groan? Truly, most gratuitously invented bugbears. I suppose if there be an " iron " law, it is that of gravitation; and if there be a physical necessity, it is that a stone, unsupported, must fall to the ground. But what is all we really know and can know about the latter phænomenon? Simply, that, in all human experience, stones have fallen to the ground under these conditions; that we have not the smallest reason for believing that any stone so circumstanced will not fall to the ground; and that we have, on the contrary, every

reason to believe that it will so fall. It is very convenient to indicate that all the conditions of belief have
been fulfilled in this case, by calling the statement that
unsupported stones will fall to the ground, " a law of
nature." But when, as commonly happens, we change
will into *must*, we introduce an idea of necessity which
most assuredly does not lie in the observed facts, and
has no warranty that I can discover elsewhere. For my
part, I utterly repudiate and anathematize the intruder.
Fact I know; and Law I know; but what is this Necessity, save an empty shadow of my own mind's
throwing?

But, if it is certain that we can have no knowledge
of the nature of either matter or spirit, and that the
notion of necessity is something illegitimately thrust
into the perfectly legitimate conception of law, the
materialistic position that there is nothing in the world
but matter, force, and necessity, is as utterly devoid of
justification as the most baseless of theological dogmas.
The fundamental doctrines of materialism, like those of
spiritualism, and most other " isms," lie outside " the
limits of philosophical inquiry," and David Hume's great
service to humanity is his irrefragable demonstration of
what these limits are. Hume called himself a sceptic,
and therefore others cannot be blamed if they apply the
same title to him; but that does not alter the fact that
the name, with its existing implications, does him gross
injustice.

If a man asks me what the politics of the inhabitants
of the moon are, and I reply that I do not know; that
neither I, nor any one else, have any means of knowing;
and that, under these circumstances, I decline to trouble

myself about the subject at all, I do not think he has any right to call me a sceptic. On the contrary, in replying thus, I conceive that I am simply honest and truthful, and show a proper regard for the economy of time. So Hume's strong and subtle intellect takes up a great many problems about which we are naturally curious, and shows us that they are essentially questions of lunar politics, in their essence incapable of being answered, and therefore not worth the attention of men who have work to do in the world. And he thus ends one of his essays :—

"If we take in hand any volume of Divinity, or school metaphysics, for instance, let us ask, *Does it contain any abstract reasoning concerning quantity or number ?* No. *Does it contain any experimental reasoning concerning matter of fact and existence ?* No. Commit it then to the flames ; for it can contain nothing but sophistry and illusion."[1]

Permit me to enforce this most wise advice. Why trouble ourselves about matters of which, however important they may be, we do know nothing, and can know nothing? We live in a world which is full of misery and ignorance, and the plain duty of each and all of us is to try to make the little corner he can influence somewhat less miserable and somewhat less ignorant than it was before he entered it. To do this effectually it is necessary to be fully possessed of only two beliefs: the first, that the order of nature is ascertainable by our faculties to an extent which is practically unlimited; the second, that our volition counts for something as a condition of the course of events.

Each of these beliefs can be verified experimentally,

[1] Hume's Essay "Of the Academical or Sceptical Philosophy," in the "Inquiry concerning the Human Understanding."

as often as we like to try.   Each, therefore, stands upon the strongest foundation upon which any belief can rest, and forms one of our highest truths.   If we find that the ascertainment of the order of nature is facilitated by using one terminology, or one set of symbols, rather than another, it is our clear duty to use the former ; and no harm can accrue, so long as we bear in mind, that we are dealing merely with terms and symbols.

In itself it is of little moment whether we express the phænomena of matter in terms of spirit ; or the phænomena of spirit, in terms of matter : matter may be regarded as a form of thought, thought may be regarded as a property of matter—each statement has a certain relative truth.   But with a view to the progress of science, the materialistic terminology is in every way to be preferred.   For it connects thought with the other phænomena of the universe, and suggests inquiry into the nature of those physical conditions, or concomitants of thought, which are more or less accessible to us, and a knowledge of which may, in future, help us to exercise the same kind of control over the world of thought, as we already possess in respect of the material world ; whereas, the alternative, or spiritualistic, terminology is utterly barren, and leads to nothing but obscurity and confusion of ideas.

Thus there can be little doubt, that the further science advances, the more extensively and consistently will all the phænomena of nature be represented by materialistic formulæ and symbols.

But the man of science, who, forgetting the limits of philosophical inquiry, slides from these formulæ and symbols into what is commonly understood by materialism, seems to me to place himself on a level with the mathematician, who should mistake the x's and y's, with which he works his problems, for real entities—and with this further disadvantage, as compared with the mathematician, that the blunders of the latter are of no practical consequence, while the errors of systematic materialism may paralyse the energies and destroy the beauty of a life.

VIII.

THE SCIENTIFIC ASPECTS OF POSITIVISM.

It is now some sixteen or seventeen years since I became acquainted with the "Philosophie Positive," the "Discours sur l'Ensemble du Positivisme," and the "Politique Positive" of Auguste Comte. I was led to study these works partly by the allusions to them in Mr. Mill's "Logic," partly by the recommendation of a distinguished theologian, and partly by the urgency of a valued friend, the late Professor Henfrey, who looked upon M. Comte's bulky volumes as a mine of wisdom, and lent them to me that I might dig and be rich. After due perusal, I found myself in a position to echo my friend's words, though I may have laid more stress on the "mine" than on the "wisdom." For I found the veins of ore few and far between, and the rock so apt to run to mud, that one incurred the risk of being intellectually smothered in the working. Still, as I was glad to acknowledge, I did come to a nugget here and there; though not, so far as my experience went, in the discussions on the philosophy of the physical sciences, but in the chapters on speculative and practical sociology. In these there was indeed much to arouse the liveliest interest in one whose boat had broken away from the old moorings, and who had been content " to lay out an anchor by the stern" until daylight should break and the fog clear.   Nothing could be more interesting to a student of biology than to see the study of the biological sciences laid down, as an essential part of the prolegomena of a new view of social phænomena. Nothing could be more satisfactory to a worshipper of the severe truthfulness of science than the attempt to dispense with all beliefs, save such as could brave the light, and seek, rather than fear, criticism ; while, to a lover of courage and outspokenness, nothing could be more touching than the placid announcement on the title-page of the "Discours sur l'Ensemble du Positivisme," that its author proposed

> " Réorganiser, sans Dieu ni roi,
> Par le culte systématique de l'Humanité,"

the shattered frame of modern society.

In those days I knew my " Faust " pretty well, and, after reading this word of might, I was minded to chant the well-known stanzas of the " Geisterchor "—

> "Weh ! Weh !
> Die schöne welt.
> Sie stürzt, sie zerfällt
> Wir tragen
> Die Trümmern ins Nichts hinüber.
> Mächtiger
> Der Erdensöhne,
> Prächtiger,
> Baue sie wieder
> In deinem Busen baue sie auf."

Great, however, was my perplexity, not to say disappointment, as I followed the progress of this "mighty son of earth" in his work of reconstruction. Undoubtedly "Dieu" disappeared, but the "Nouveau Grand-Être Suprême," a gigantic fetish, turned out brannew by M. Comte's own hands, reigned in his stead. "Roi" also was not heard of; but, in his place, I found a minutely-defined social organization, which, if it ever came into practice, would exert a despotic authority such as no sultan has rivalled, and no Puritan presbytery, in its palmiest days, could hope to excel. While, as for the "culte systématique de l'Humanité," I, in my blindness, could not distinguish it from sheer Popery, with M. Comte in the chair of St. Peter, and the names of most of the saints changed. To quote "Faust" again, I found myself saying with Gretchen,—

> "Ungefähr sagt das der Pfarrer auch
> Nur mit ein bischen andern Worten."

Rightly or wrongly, this was the impression which, all those years ago, the study of M. Comte's works left on my mind, combined with the conviction, which I shall always be thankful to him for awakening in me, that the organization of society upon a new and purely scientific basis is not only practicable, but is the only political object much worth fighting for.

As I have said, that part of M. Comte's writings which deals with the philosophy of physical science appeared to me to possess singularly little value, and to show that he had but the most superficial, and merely second-hand, knowledge of most branches of what is usually understood by science. I do not mean by this merely to say that Comte was behind our present knowledge, or that he was unacquainted with the details of

the science of his own day.   No one could justly make
such defects cause of complaint in a philosophical writer
of the past generation.   What struck me was his want of
apprehension of the great features of science ; his strange
mistakes as to the merits of his scientific contemporaries;
and his ludicrously erroneous notions about the part which
some of the scientific doctrines current in his time were
destined to play in the future.   With these impressions
in my mind, no one will be surprised if I acknowledge
that, for these sixteen years, it has been a periodical
source of irritation to me to find M. Comte put forward
as a representative of scientific thought ; and to observe
that writers whose philosophy had its legitimate parent
in Hume, or in themselves, were labelled " Comtists " or
" Positivists " by public writers, even in spite of vehe-
ment protests to the contrary.   It has cost Mr. Mill
hard rubbings to get that label off ; and I watch Mr.
Spencer, as one regards a good man struggling with
adversity, still engaged in eluding its adhesiveness, and
ready to tear away skin and all, rather than let it stick.
My own turn might come next ; and, therefore, when
an eminent prelate the other day gave currency and
authority to the popular confusion, I took an opportunity of incidentally revindicating Hume's property in
the so-called " New Philosophy," and, at the same time,
of repudiating Comtism on my own behalf.[1]

---

[1] I am glad to observe that Mr. Congreve, in the criticism with which he
has favoured me in the number of the *Fortnightly Review* for April 1869, does
not venture to challenge the justice of the claim I make for Hume.   He merely
suggests that I have been wanting in candour in not mentioning Comte's high
opinion of Hume.   After mature reflection I am unable to discern my fault.
If I had suggested that Comte had borrowed from Hume without acknowledgment; or if, instead of trying to express my own sense of Hume's merits with

The few lines devoted to Comtism in my paper on the "Physical Basis of Life" were, in intention, strictly limited to these two purposes. But they seem to have given more umbrage than I intended they should, to the followers of M. Comte in this country, for some of whom, let me observe in passing, I entertain a most unfeigned respect ; and Mr. Congreve's recent article gives expression to the displeasure which I have excited among the members of the Comtian body.

Mr. Congreve, in a peroration which seems especially intended to catch the attention of his readers, indignantly challenges me to admire M. Comte's life, " to deny that it has a marked character of grandeur about it ; " and he uses some very strong language because I show no sign of veneration for his idol. I confess I do not care to occupy myself with the denigration of a man who, on the whole, deserves to be spoken of with respect. Therefore, I shall enter into no statement of the reasons

the modesty which becomes a writer who has no authority in matters of philosophy, I had affirmed that no one had properly appreciated him, Mr. Congreve's remarks would apply : but as I did neither of these things, they appear to me to be irrelevant, if not unjustifiable. And even had it occurred to me to quote M. Comte's expressions about Hume, I do not know that I should have cited them, inasmuch as, on his own showing, M. Comte occasionally speaks very decidedly touching writers of whose works he has not read a line. Thus, in Tome VI. of the "Philosophie Positive," p. 619, M. Comte writes : "Le plus grand des métaphysiciens modernes, l'illustre Kant, a noblement mérité une éternelle admiration en tentant, le premier, d'échapper directement à l'absolu philosophique par sa célèbre conception de la double réalité, à la fois objective et subjective, qui indique un si juste sentiment de la saine philosophie."

But in the "Préface Personnelle" in the same volume, p. 35, M. Comte tells us :—" Je n'ai jamais lu, en aucune langue, ni Vico, *ni Kant*, ni Herder, ni Hegel, &c. ; je ne connais leurs divers ouvrages que d'après quelques relations indirectes et certains extraits fort insuffisants."

Who knows but that the " &c." may include Hume ? And in that case what is the value of M. Comte's praise of him ?

which lead me unhesitatingly to accept Mr. Congreve's
challenge, and to refuse to recognise anything which deserves the name of grandeur of character in M. Comte,
unless it be his arrogance, which is undoubtedly sublime.
All I have to observe is, that if Mr. Congreve is justified
in saying that I speak with a tinge of contempt for his
spiritual father, the reason for such colouring of my
language is to be found in the fact, that, when I wrote,
I had but just arisen from the perusal of a work with
which he is doubtless well acquainted, M. Littré's
"Auguste Comte et la Philosophie Positive."

Though there are tolerably fixed standards of right
and wrong, and even of generosity and meanness, it
may be said that the beauty, or grandeur, of a life is
more or less a matter of taste; and Mr. Congreve's
notions of literary excellence are so different from mine
that, it may be, we should diverge as widely in our
judgment of moral beauty or ugliness. Therefore, while
retaining my own notions, I do not presume to quarrel
with his. But when Mr. Congreve devotes a great deal
of laboriously guarded insinuation to the endeavour to
lead the public to believe that I have been guilty of the
dishonesty of having criticised Comte without having
read him, I must be permitted to remind him that he
has neglected the well-known maxim of a diplomatic
sage, "If you want to damage a man, you should say
what is probable, as well as what is true."

And when Mr. Congreve speaks of my having an advantage over him in my introduction of "Christianity"
into the phrase that "M. Comte's philosophy, in practice,
might be described as Catholicism *minus* Christianity;"
intending thereby to suggest that I have, by so doing,

desired to profit by an appeal to the *odium theologicum,* —he lays himself open to a very unpleasant retort.

What if I were to suggest that Mr. Congreve had not read Comte's works ; and that the phrase " the context shows that the view of the writer ranges—however superficially—over the whole works. This is obvious from the mention of Catholicism," demonstrates that Mr. Congreve has no acquaintance with the " Philosophie Positive " ? I think the suggestion would be very unjust and unmannerly, and I shall not make it.   But the fact remains, that this little epigram of mine, which has so greatly provoked Mr. Congreve, is neither more nor less than a condensed paraphrase of the following passage, which is to be found at page 344 of the fifth volume of the " Philosophie Positive :" [1]—

" La seule solution possible de ce grand problème historique, qui n'a jamais pu être philosophiquement posé jusqu'ici, consiste à concevoir, en sens radicalement inverse des notions habituelles, *que ce qui devait nécessairement périr ainsi, dans le catholicisme, c'était la doctrine, et non l'organisation,* qui n'a été passagèrement ruinée que par suite de son inévitable adhérence élémentaire à la philosophie théologique, destinée à succomber graduellement sous l'irrésistible émancipation de la raison humaine ; *tandis qu'une telle constitution, convenablement reconstruite sur des bases intellectuelles à la fois plus étendues et plus stables, devra finalement présider à l'indispensable réorganisation spirituelle des sociétés modernes, sauf les différences essentielles spontanément correspondantes à l'extrême diversité des doctrines fondamentales ;* à moins de supposer, ce qui serait certainement contradictoire à l'ensemble des lois de notre nature, que les immenses efforts de tant de grands hommes, secondés par la persévérante sollicitude des nations civilisées, dans la fondation séculaire de ce chef-d'œuvre politique de la sagesse humaine, doivent être enfin irrévocablement perdus pour l'élite de l'humanité sauf les résultats, capitaux mais provisoires, qui s'y rapportaient immédiatement.   Cette explication générale, déjà évidemment motivée par la suite des considérations propres à ce chapitre,

---

[1] Now and always I quote the second edition, by Littré.

sera de plus en plus confirmée par tout le reste de notre opération historique, *dont elle constituera spontanément la principale conclusion politique.*"

Nothing can be clearer. Comte's ideal, as stated by himself, is Catholic organization without Catholic doctrine, or, in other words, Catholicism *minus* Christianity. Surely it is utterly unjustifiable to ascribe to me base motives for stating a man's doctrines, as nearly as may be, in his own words!

My readers would hardly be interested were I to follow Mr. Congreve any further, or I might point out that the fact of his not having heard me lecture is hardly a safe ground for his speculations as to what I do not teach. Nor do I feel called upon to give any opinion as to M. Comte's merits or demerits as regards sociology. Mr. Mill (whose competence to speak on these matters I suppose will not be questioned, even by Mr. Congreve) has dealt with M. Comte's philosophy from this point of view, with a vigour and authority to which I cannot for a moment aspire; and with a severity, not unfrequently amounting to contempt, which I have not the wish, if I had the power, to surpass. I, as a mere student in these questions, am content to abide by Mr. Mill's judgment until some one shows cause for its reversal, and I decline to enter into a discussion which I have not provoked.

The sole obligation which lies upon me is to justify so much as still remains without justification of what I have written respecting Positivism—namely, the opinion expressed in the following paragraph:—

" In so far as my study of what specially characterises the Positive Philosophy has led me, I find therein little or nothing of any scientific value, and a great deal which is as thoroughly antagonistic to the very essence of science as anything in ultramontane Catholicism."

Here are two propositions : the first, that the " Philosophie Positive " contains little or nothing of any scientific value ; the second, that Comtism is, in spirit, anti-scientific. I shall endeavour to bring forward ample evidence in support of both.

I. No one who possesses even a superficial acquaintance with physical science can read Comte's " Leçons " without becoming aware that he was at once singularly devoid of real knowledge on these subjects, and singularly unlucky. What is to be thought of the contemporary of Young and of Fresnel, who never misses an opportunity of casting scorn upon the hypothesis of an ether—the fundamental basis not only of the undulatory theory of light, but of so much else in modern physics—and whose contempt for the intellects of some of the strongest men of his generation was such, that he puts forward the mere existence of night as a refutation of the undulatory theory ? [1] What a wonderful gauge of his own value as a scientific critic does he afford, by whom we are informed that phrenology is a great science, and psychology a chimæra ; that Gall was one of the great men of his age, and that Cuvier was " brilliant but superficial " ! [2] How unlucky must one consider the bold speculator who, just before the dawn of modern histology—which is simply the application of the microscope to anatomy—reproves what he calls " the abuse of microscopic investigations," and " the exaggerated credit " attached to them ; who, when the morphological uniformity of the tissues of the great majority of plants and animals was on the eve of being

---

[1] " Philosophie Positive," ii. p. 440.
[2] " Le brillant mais superficiel Cuvier."—*Philosophie Positive*, vi. p. 383.

demonstrated, treated with ridicule those who attempt
to refer all tissues to a " tissu générateur," formed by
" le chimérique et inintelligible assemblage d'une sorte
de monades organiques, qui seraient dès lors les vrais
éléments primordiaux de tout corps vivant ; " [1] and who
finally tells us, that all the objections against a linear
arrangement of the species of living beings are in their
essence foolish, and that the order of the animal series is
" necessarily linear," [2] when the exact contrary is one of
the best-established and the most important truths of
zoology.   Appeal to mathematicians, astronomers, physicists,[3] chemists, biologists, about the " Philosophie Positive," and they all, with one consent, begin to make
protestation that, whatever M. Comte's other merits, he
has shed no light upon the philosophy of their particular
studies.

To be just, however, it must be admitted that even
M. Comte's most ardent disciples are content to be
judiciously silent about his knowledge or appreciation of
the sciences themselves, and prefer to base their master's
claims to scientific authority upon his " law of the
three states," and his " classification of the sciences."
But here, also, I must join issue with them as completely
as others—notably Mr. Herbert Spencer—have done
before me.   A critical examination of what M. Comte
has to say about the " law of the three states " brings out
nothing but a series of more or less contradictory state-

---

[1] " Philosophie Positive," iii. p. 369.   [2] Ibid. p. 387.

[3] Hear the late Dr. Whewell, who calls Comte " a shallow pretender," so
far as all the modern sciences, except astronomy, are concerned ; and tells us
that "his pretensions to discoveries are, as Sir John Herschel has shown,
absurdly fallacious."—" Comte and Positivism," *Macmillan's Magazine*,
March 1866.

ments of an imperfectly apprehended truth; and his "classification of the sciences," whether regarded historically or logically, is, in my judgment, absolutely worthless.

Let us consider the law of " the three states" as it is put before us in the opening of the first Leçon of the "Philosophie Positive:"—

" En étudiant ainsi le développement total de l'intelligence humaine dans ses diverses sphères d'activité, depuis son premier essor le plus simple jusqu'à nos jours, je crois avoir découvert une grande loi fondamentale, à laquelle il est assujetti par une nécessité invariable, et qui me semble pouvoir être solidement établie, soit sur les preuves rationelles fournies par la connaissance de notre organisation, soit sur les vérifications historiques résultant d'un examen attentif du passé. Cette loi consiste en ce que chacune de nos conceptions principales, chaque branche de nos connaissances, passe successivement par trois états théoriques différents; l'état théologique, ou fictif; l'état métaphysique, ou abstrait; l'état scientifique, ou positif.   En d'autres termes, l'esprit humain, par sa nature, emploie successivement dans chacune de ses recherches trois méthodes de philosopher, dont *le caractère est essentiellement différent et même radicalement* opposé; d'abord la méthode théologique, ensuite la méthode métaphysique, et enfin la méthode positive.   De là, trois sortes de philosophie, ou de systèmes généraux de conceptions sur l'ensemble des phénomènes *qui s'excluent mutuellement;* la première est le point de départ nécessaire de l'intelligence humaine; la troisième, son état fixe et définitif; la seconde est uniquement destinée à servir de transition."[1]

Nothing can be more precise than these statements, which may be put into the following propositions:—

(*a*) The human intellect is subjected to the law by an invariable necessity, which is demonstrable, *à priori,* from the nature and constitution of the intellect; while, as a matter of historical fact, the human intellect has been subjected to the law.

(*b*) Every branch of human knowledge passes through

[1] "Philosophie Positive," i. pp. 8, 9.

the three states, necessarily beginning with the first stage.

(*c*) The three states mutually exclude one another, being essentially different, and even radically opposed.

Two questions present themselves. Is M. Comte consistent with himself in making these assertions? And is he consistent with fact? I reply to both questions in the negative; and, as regards the first, I bring forward as my witness a remarkable passage which is to be found in the fourth volume of the " Philosophie Positive " (p. 491), when M. Comte had had time to think out, a little more fully, the notions crudely stated in the first volume :—

" A proprement parler, la philosophie théologique, même dans notre première enfance, individuelle ou sociale, n'a jamais pu être rigoureusement universelle, c'est-à-dire que, pour les ordres quelconques de phénomènes, *les faits les plus simples et les plus communs ont toujours été regardés comme essentiellement assujettis à des lois naturelles, au lieu d'être attribués à l'arbitraire volonté des agents surnaturels.* L'illustre Adam Smith a, par exemple, très-heureusement remarqué dans ses essais philosophiques, qu'on ne trouvait, en aucun temps ni en aucun pays, un dieu pour la pesanteur. *Il en est ainsi, en général, même à l'égard des sujets les plus compliqués, envers tous les phénomènes assez élémentaires et assez familiers pour que la parfaite invariabilité de leurs relations effectives ait toujours dû frapper spontanément l'observateur le moins préparé.* Dans l'ordre moral et social, qu'une vaine opposition voudrait aujourd'hui systématiquement interdire à la philosophie positive, il y a eu nécessairement, en tout temps, la pensée des lois naturelles, relativement aux plus simples phénomènes de la vie journalière, comme l'exige évidemment la conduite générale de notre existence réelle, individuelle ou sociale, qui n'aurait pu jamais comporter aucune prévoyance quelconque, si tous les phénomènes humains avaient été rigoureusement attribués à des agents surnaturels, puisque dès lors la prière aurait logiquement constitué la seule ressource imaginable pour influer sur le cours habituel des actions humaines. *On doit même remarquer, à ce sujet, que c'est, au contraire,*

*l'ébauche spontanée des premières lois naturelles propres aux actes indi-
viduels ou sociaux qui, fictivement transportée à tous les phénomènes du
monde extérieur, a d'abord fourni, d'après nos explications précédentes, le
vrai principe fondamental de la philosophie théologique. Ainsi, le germe
élémentaire de la philosophie positive est certainement tout aussi primitif
au fond que celui de la philosophie théologique elle-même, quoi qu'il n'ait
pu se développer que beaucoup plus tard.* Une telle notion importe
extrêmement à la parfaite rationalité de notre théorie sociologique, puisque la vie humaine ne pouvant jamais offrir aucune véritable création
quelconque, mais toujours une simple évolution graduelle, l'essor final
de l'esprit positif deviendrait scientifiquement incompréhensible, si,
dès l'origine, on n'en concevait, à tous égards, les premiers rudiments
nécessaires. Depuis cette situation primitive, à mesure que nos
observations se sont spontanément étendues et généralisées, cet essor,
d'abord à peine appréciable, a constamment suivi, sans cesser long-
temps d'être subalterne, une progression très-lente, mais continue, la
philosophie théologique restant toujours réservée pour les phénomènes,
de moins en moins nombreux, dont les lois naturelles ne pouvaient
encore être aucunement connues."

Compare the propositions implicitly laid down here with those contained in the earlier volume. (*a*) As a matter of fact, the human intellect has *not* been invariably subjected to the law of the three states, and therefore the necessity of the law *cannot* be demonstrable *à priori*. (*b*) Much of our knowledge of all kinds has *not* passed through the three states, and more particularly, as M. Comte is careful to point out, not through the first. (*c*) The positive state has more or less co-existed with the theological, from the dawn of human intelligence. And, by way of completing the series of contradictions, the assertion that the three states are "essentially different and even radically opposed," is met a little lower on the same page by the declaration that "the metaphysical state is, at bottom, nothing but a simple general modification of the first;" while, in the fortieth Leçon, as also in the

interesting early essay entitled " Considérations philosophiques sur les Sciences et les Savants (1825)," the
three states are practically reduced to two. " Le véritable esprit général de toute philosophie théologique
ou métaphysique consiste à prendre pour principe, dans
l'explication des phénomènes du monde extérieur, notre
sentiment immédiat des phénomènes humains ; tandis
que au contraire, la philosophie positive est toujours
caractérisée, non moins profondément, par la subordination nécessaire et rationnelle de la conception de l'homme
à celle du monde."[1]

I leave M. Comte's disciples to settle which of these
contradictory statements expresses their master's real
meaning. All I beg leave to remark is, that men of
science are not in the habit of paying much attention
to "laws" stated in this fashion.

The second statement is undoubtedly far more rational
and consistent with fact than the first; but I cannot
think it is a just or adequate account of the growth
of intelligence, either in the individual man, or in the
human species. Any one who will carefully watch the
development of the intellect of a child will perceive
that, from the first, its mind is mirroring nature in two
different ways. On the one hand, it is merely drinking
in sensations and building up associations, while it forms
conceptions of things and their relations which are more
thoroughly " positive," or devoid of entanglement with
hypotheses of any kind, than they will ever be in after-
life. No child has recourse to imaginary personifications
in order to account for the ordinary properties of objects
which are not alive, or do not represent living things. It

[1] "Philosophie Positive," iii. p. 188.

does not imagine that the taste of sugar is brought about by a god of sweetness, or that a spirit of jumping causes a ball to bound. Such phænomena, which form the basis of a very large part of its ideas, are taken as matters of course—as ultimate facts which suggest no difficulty and need no explanation. So far as all these common, though important, phænomena are concerned, the child's mind is in what M. Comte would call the "positive" state.

But, side by side with this mental condition, there rises another. The child becomes aware of itself as a source of action and a subject of passion and of thought. The acts which follow upon its own desires are among the most interesting and prominent of surrounding occurrences ; and these acts, again, plainly arise either out of affections caused by surrounding things, or of other changes in itself. Among these surrounding things, the most interesting and important are mother and father, brethren and nurses. The hypothesis that these wonderful creatures are of like nature to itself is speedily forced upon the child's mind ; and this primitive piece of anthropomorphism turns out to be a highly successful speculation, which finds its justification at every turn. No wonder, then, that it is extended to other similarly interesting objects which are not too unlike these—to the dog, the cat, and the canary, the doll, the toy, and the picture-book—that these are endowed with wills and affections, and with capacities for being "good" and "naughty." But surely it would be a mere perversion of language to call this a "theological" state of mind, either in the proper sense of the word "theological," or as contrasted with "scientific" or "positive." The child does not worship either father or mother, dog or doll. On the contrary, nothing is more curious than the absolute irreverence, if I may so say, of a kindly-treated young child; its tendency to believe in itself as the centre of the universe, and its disposition to exercise despotic tyranny over those who could crush it with a finger.

Still less is there anything unscientific, or anti-scientific, in this infantile anthropomorphism. The child observes that many phænomena are the consequences of affections of itself; it soon has excellent reasons for the belief that many other phænomena are consequences of the affections of other beings, more or less like itself. And having thus good evidence for believing that many of the most interesting occurrences about it are explicable on the hypothesis that they are the work of intelligences like itself—having discovered a *vera causa* for many phænomena—why should the child limit the application of so fruitful an hypothesis? The dog has a sort of intelligence, so has the cat; why should not the doll and the picture-book also have a share, proportioned to their likeness to intelligent things?

The only limit which does arise is exactly that which, as a matter of science, should arise; that is to say, the anthropomorphic interpretation is applied only to those phænomena which, in their general nature, or their apparent capriciousness, resemble those which the child observes to be caused by itself, or by beings like itself. All the rest are regarded as things which explain themselves, or are inexplicable.

It is only at a later stage of intellectual development that the intelligence of man awakes to the apparent conflict between the anthropomorphic, and what I may

call the physical,[1] aspect of nature, and either endeavours
to extend the anthropomorphic view over the whole of
nature—which is the tendency of theology ; or to give
the same exclusive predominance to the physical view—
which is the tendency of science ; or adopts a middle
course, and taking from the anthropomorphic view its
tendency to personify, and from the physical view its
tendency to exclude volition and affection, ends in what
M. Comte calls the "metaphysical" state—"metaphysical," in M. Comte's writings, being a general term of
abuse for anything he does not like.

What is true of the individual is, *mutatis mutandis,*
true of the intellectual development of the species. It
is absurd to say of men in a state of primitive savagery,
that all their conceptions are in a theological state.
Nine-tenths of them are eminently realistic, and as
"positive" as ignorance and narrowness can make them.
It no more occurs to a savage than it does to a child,
to ask the why of the daily and ordinary occurrences
which form the greater part of his mental life. But
in regard to the more striking, or out-of-the-way, events,
which force him to speculate, he is highly anthropomorphic ; and, as compared with a child, his anthropomorphism is complicated by the intense impression
which the death of his own kind makes upon him,
as indeed it well may. The warrior, full of ferocious

---

[1] The word "positive" is in every way objectionable. In one sense it
suggests that mental quality which was undoubtedly largely developed in
M. Comte, but can best be dispensed with in a philosopher ; in another, it is
unfortunate in its application to a system which starts with enormous nega-
tions ; in its third, and specially philosophical sense, as implying a system of
thought which assumes nothing beyond the content of observed facts, it
implies that which never did exist, and never will.

energy, perhaps the despotic chief of his tribe, is
suddenly struck down. A child may insult the man
a moment before so awful; a fly rests, undisturbed, on
the lips from which undisputed command issued. And
yet the bodily aspect of the man seems hardly more
altered than when he slept, and, sleeping, seemed to
himself to leave his body and wander through dream-
land. What then if that something, which is the essence
of the man, has really been made to wander by the
violence done to it, and is unable, or has forgotten,
to come back to its shell? Will it not retain somewhat of the powers it possessed during life? May
it not help us if it be pleased, or (as seems to be
by far the more general impression) hurt us if it be
angered? Will it not be well to do towards it those
things which would have soothed the man and put
him in good humour during his life? It is impossible
to study trustworthy accounts of savage thought without seeing, that some such train of ideas as this, lies at
the bottom of their speculative beliefs.

There are savages without God, in any proper sense
of the word, but none without ghosts. And the Fetishism, Ancestor-worship, Hero-worship, and Demonology
of primitive savages, are all, I believe, different manners
of expression of their belief in ghosts, and of the
anthropomorphic interpretation of out-of-the-way events,
which is its concomitant. Witchcraft and sorcery are
the practical expressions of these beliefs; and they
stand in the same relation to religious worship as the
simple anthropomorphism of children, or savages, does
to theology.

In the progress of the species from savagery to

advanced civilization, anthropomorphism grows into theology, while physicism (if I may so call it) develops into science ; but the development of the two is contemporaneous, not successive.   For each, there long exists an assured province which is not invaded by the other ; while, between the two, lies a debateable land, ruled by a sort of bastards, who owe their complexion to physicism and their substance to anthropomorphism, and are M. Comte's particular aversions—metaphysical entities.

But, as the ages lengthen, the borders of Physicism increase.   The territories of the bastards are all annexed to science ; and even Theology, in her purer forms, has ceased to be anthropomorphic, however she may talk.   Anthropomorphism has taken stand in its last fortress—man himself.   But science closely invests the walls ; and Philosophers gird themselves for battle upon the last and greatest of all speculative problems— Does human nature possess any free, volitional, or truly anthropomorphic element, or is it only the cunningest of all Nature's clocks ?   Some, among whom I count myself, think that the battle will for ever remain a drawn one, and that, for all practical purposes, this result is as good as anthropomorphism winning the day.

The classification of the sciences, which, in the eyes of M. Comte's adherents, constitutes his second great claim to the dignity of a scientific philosopher, appears to me to be open to just the same objections as the law of the three states.   It is inconsistent in itself, and it is inconsistent with fact.   Let us consider the main points of this classification successively :—

" Il faut distinguer par rapport à tous les ordres des phénomènes, deux genres de sciences naturelles ; les unes abstraites, générales, ont pour objet la découverte des lois qui régissent les diverses classes de phénomènes, en considérant tous les cas qu'on peut concevoir ; les autres concrètes, particulières, descriptives, et qu'on désigne quelquefois sous le nom des sciences naturelles proprement dites, consistent dans l'application de ces lois à l'histoire effective des différents êtres existants." [1]

The " abstract" sciences are subsequently said to be mathematics, astronomy, physics, chemistry, physiology, and social physics—the titles of the two latter being subsequently changed to biology and sociology. M. Comte exemplifies the distinction between his abstract and his concrete sciences as follows :—

" On pourra d'abord l'apercevoir très-nettement en comparant, d'une part, la physiologie générale, et d'une autre part la zoologie et la botanique proprement dites. Ce sont évidemment, en effet, deux travaux d'un caractère fort distinct, que d'étudier, en général, les lois de la vie, ou de déterminer le mode d'existence de chaque corps vivant, en particulier. *Cette seconde étude, en outre, est nécessairememt fondée sur la première.*"—P. 57.

All the unreality and mere bookishness of M. Comte's knowledge of physical science comes out in the passage I have italicised. " The special study of living beings is based upon a general study of the laws of life !" What little I know about the matter leads me to think, that, if M. Comte had possessed the slightest practical acquaintance with biological science, he would have turned his phraseology upside down, and have perceived that we can have no knowledge of the general laws of life, except that which is based upon the study of particular living beings.

The illustration is surely unluckily chosen ; but the language in which these so-called abstract sciences are

[1] " Philosophie Positive," i. p. 56.

defined seems to me to be still more open to criticism. With what propriety can astronomy, or physics, or chemistry, or biology, be said to occupy themselves with the consideration of "all conceivable cases" which fall within their respective provinces? Does the astronomer occupy himself with any other system of the universe than that which is visible to him? Does he speculate upon the possible movements of bodies which may attract one another in the inverse proportion of the cube of their distances, say? Does biology, whether "abstract" or "concrete," occupy itself with any other form of life than those which exist, or have existed? And, if the abstract sciences embrace all conceivable cases of the operation of the laws with which they are concerned, would not they, necessarily, embrace the subjects of the concrete sciences, which, inasmuch as they exist, must needs be conceivable? In fact, no such distinction as that which M. Comte draws is tenable. The first stage of his classification breaks by its own weight.

But granting M. Comte his six abstract sciences, he proceeds to arrange them according to what he calls their natural order or hierarchy, their places in this hierarchy being determined by the degree of generality and simplicity of the conceptions with which they deal. Mathematics occupies the first, astronomy the second, physics the third, chemistry the fourth, biology the fifth, and sociology the sixth and last place in the series. M. Comte's arguments in favour of this classification are first—

"Sa conformité essentielle avec la co-ordination, en quelque sorte spontanée, qui se trouve en effet implicitement admise par les savants livrés à l'étude des diverse branches de la philosophie naturelle."

But I absolutely deny the existence of this conformity. If there is one thing clear about the progress of modern science, it is the tendency to reduce all scientific problems, except those which are purely mathematical, to questions of molecular physics—that is to say, to the attractions, repulsions, motions, and co-ordination of the ultimate particles of matter. Social phænomena are the result of the interaction of the components of society, or men, with one another and the surrounding universe. But, in the language of physical science, which, by the nature of the case, is materialistic, the actions of men, so far as they are recognisable by science, are the results of molecular changes in the matter of which they are composed; and, in the long run, these must come into the hands of the physicist. *A fortiori,* the phænomena of biology and of chemistry are, in their ultimate analysis, questions of molecular physics. Indeed, the fact is acknowledged by all chemists and biologists who look beyond their immediate occupations. And it is to be observed, that the phænomena of biology are as directly and immediately connected with molecular physics as are those of chemistry. Molar physics, chemistry, and biology are not three successive steps in the ladder of knowledge, as M. Comte would have us believe, but three branches springing from the common stem of molecular physics.

As to astronomy, I am at a loss to understand how any one who will give a moment's attention to the nature of the science can fail to see that it consists of two parts: first, of a description of the phænomena, which is as much entitled as descriptive zoology, or botany, is, to the name of natural history; and, secondly, of an explanation of the phænomena, furnished by the laws of a force—gravitation—the study of which is as much a part of physics, as is that of heat, or electricity. It would be just as reasonable to make the study of the heat of the sun a science preliminary to the rest of thermotics, as to place the study of the attraction of the bodies, which compose the universe in general, before that of the particular terrestrial bodies, which alone we can experimentally know. Astronomy, in fact, owes its perfection to the circumstance that it is the only branch of natural history, the phænomena of which are largely expressible by mathematical conceptions, and which can be, to a great extent, explained by the application of very simple physical laws.

With regard to mathematics, it is to be observed, in the first place, that M. Comte mixes up under that head the pure relations of space and of quantity, which are properly included under the name, with rational mechanics and statics, which are mathematical developments of the most general conceptions of physics, namely, the notions of force and of motion. Relegating these to their proper place in physics, we have left pure mathematics, which can stand neither at the head, nor at the tail, of any hierarchy of the sciences, since, like logic, it is equally related to all; though the enormous practical difficulty of applying mathematics to the more complex phænomena of nature removes them, for the present, out of its sphere.

On this subject of mathematics, again, M. Comte indulges in assertions which can only be accounted for by his total ignorance of physical science practically. As for example :—

"C'est donc par l'étude des mathématiques, *et seulement par elle,* que l'on peut se faire une idée juste et approfondie de ce que c'est qu'une *science.* C'est là *uniquement* qu'on doit chercher à connaître avec précision *la méthode générale que l'esprit humain emploie constamment dans toutes ses recherches positives,* parce que nulle part ailleurs les questions ne sont résolues d'une manière aussi complète et les déductions prolongées aussi loin avec une sévérité rigoureuse. C'est là également que notre entendement a donné les plus grandes preuves de sa force, parce que les idées qu'il y considère sont du plus haut degré d'abstraction possible dans l'ordre positif. *Toute éducation scientifique qui ne commence point par une telle étude pèche donc nécessairement par sa base.*" [1]

That is to say, the only study which can confer " a just and comprehensive idea of what is meant by science," and, at the same time, furnish an exact conception of the general method of scientific investigation, is that which knows nothing of observation, nothing of experiment, nothing of induction, nothing of causation! And education, the whole secret of which consists in proceeding from the easy to the difficult, the concrete to the abstract, ought to be turned the other way, and pass from the abstract to the concrete.

M. Comte puts a second argument in favour of his hierarchy of the sciences thus :—

"Un second caractère très-essentiel de notre classification, c'est d'être nécessairement conforme à l'ordre effectif du développement de la philosophie naturelle. C'est ce que vérifie tout ce qu'on sait de l'histoire des sciences." [2]

But Mr. Spencer has so thoroughly and completely demonstrated the absence of any correspondence between the historical development of the sciences, and their position in the Comtean hierarchy, in his essay on the

[1] " Philosophie Positive," i. p. 99.   [2] Ibid., i. p. 77.

"Genesis of Science," that I shall not waste time in repeating his refutation.

A third proposition in support of the Comtean classification of the sciences stands as follows :—

"En troisième lieu cette classification présente la propriété très-remarquable de marquer exactement la perfection relative des différentes sciences, laquelle consiste essentiellement dans le degré de précision des connaissances et dans leur co-ordination plus ou moins intime."[1]

I am quite unable to understand the distinction which M. Comte endeavours to draw in this passage in spite of his amplifications further on. Every science must consist of precise knowledge, and that knowledge must be co-ordinated into general proportions, or it is not science. When M. Comte, in exemplification of the statement I have cited, says that "les phénomènes organiques ne comportent qu'une étude à la fois moins exacte et moins systématique que les phénomènes des corps bruts," I am at a loss to comprehend what he means. If I affirm that "when a motor nerve is irritated, the muscle connected with it becomes simultaneously shorter and thicker, without changing its volume," it appears to me that the statement is as precise or exact (and not merely as true) as that of the physicist who should say, that "when a piece of iron is heated, it becomes simultaneously longer and thicker and increases in volume;" nor can I discover any difference, in point of precision, between the statement of the morphological law that "animals which suckle their young have two occipital condyles," and the enunciation of the physical

[1] "Philosophie Positive," i. p. 78.

law that "water subjected to electrolysis is replaced by an equal weight of the gases, oxygen and hydrogen." As for anatomical or physiological investigation being less "systematic" than that of the physicist or chemist, the assertion is simply unaccountable. The methods of physical science are everywhere the same in principle, and the physiological investigator who was not "systematic" would, on the whole, break down rather sooner than the inquirer into simpler subjects.

Thus M. Comte's classification of the sciences, under all its aspects, appears to me to be a complete failure. It is impossible, in an article which is already too long, to inquire how it may be replaced by a better; and it is the less necessary to do so, as a second edition of Mr. Spencer's remarkable essay on this subject has just been published. After wading through pages of the long-winded confusion and second-hand information of the "Philosophie Positive," at the risk of a *crise cérébrale*— it is as good as a shower-bath to turn to the "Classification of the Sciences," and refresh oneself with Mr. Spencer's profound thought, precise knowledge, and clear language.

II. The second proposition to which I have committed myself, in the paper to which I have been obliged to refer so often, is, that the "Positive Philosophy" contains "a great deal which is as thoroughly antagonistic to the very essence of science as is anything in ultramontane Catholicism."

What I refer to in these words, is, on the one hand, the dogmatism and narrowness which so often mark M. Comte's discussion of doctrines which he does not like, and reduce his expressions of opinion to mere

passionate puerilities ; as, for example, when he is
arguing against the assumption of an ether, or when
he is talking (I cannot call it arguing) against psychology, or political economy.   On the other hand, I allude
to the spirit of meddling systematization and regulation
which animates even the " Philosophie Positive," and
breaks out, in the latter volumes of that work, into no
uncertain foreshadowing of the anti-scientific monstro-
sities of Comte's later writings.

Those who try to draw a line of demarcation between
the spirit of the " Philosophie Positive," and that of
the " Politique " and its successors, (if I may express
an opinion from fragmentary knowledge of these last,)
must have overlooked, or forgotten, what Comte himself
labours to show, and indeed succeeds in proving, in
the " Appendice Général " of the " Politique Positive."
"Dès mon début," he writes, " je tentai de fonder le
nouveau pouvoir spirituel que j'institue aujourd'hui."
" Ma politique, loin d'être aucunement opposée à ma
philosophie, en constitue tellement la suite naturelle
que celle-ci fut directement instituée pour servir de base
à celle-là, comme le prouve cet appendice."[1]

This is quite true.   In the remarkable essay entitled
" Considérations sur le Pouvoir spirituel," published in
March 1826, Comte advocates the establishment of a
" modern spiritual power," which, he anticipates, may
exercise an even greater influence over temporal affairs,
than did the Catholic clergy, at the height of their
vigour and independence, in the twelfth century.   This
spiritual power is, in fact, to govern opinion, and to
have the supreme control over education, in each

---

[1] Loc. cit.. Préface Spéciale, pp. i. ii.

nation of the West ; and the spiritual powers of the
several European peoples are to be associated together
and placed under a common direction or " souveraineté
spirituelle."

A system of " Catholicism *minus* Christianity " was
therefore completely organized in Comte's mind, four
years before the first volume of the " Philosophie
Positive " was written ; and, naturally, the papal spirit
shows itself in that work, not only in the ways I
have already mentioned, but, notably, in the attack
on liberty of conscience which breaks out in the fourth
volume :—

"Il n'y a point de liberté de conscience en astronomie, en physique,
en chimie, en physiologie même, en ce sens que chacun trouverait
absurde de ne pas croire de confiance aux principes établis dans les
sciences par les hommes compétents."

"Nothing in ultramontane Catholicism" can, in my
judgment, be more completely sacerdotal, more entirely
anti-scientific, than this dictum. All the great steps in
the advancement of science have been made by just
those men who have not hesitated to doubt the " principles established in the sciences by competent persons ; "
and the great teaching of science—the great use of it as
an instrument of mental discipline—is its constant incul-
cation of the maxim, that the sole ground on which any
statement has a right to be believed is the impossibility
of refuting it.

Thus, without travelling beyond the limits of the
" Philosophie Positive," we find its author contemplating the establishment of a system of society, in which
an organized spiritual power shall over-ride and direct
the temporal power, as completely as the Innocents and

Gregorys tried to govern Europe in the middle ages ; and repudiating the exercise of liberty of conscience against the "*hommes compétents,*" of whom, by the assumption, the new priesthood would be composed. Was Mr. Congreve as forgetful of this, as he seems to have been of some other parts of the " Philosophie Positive," when he wrote, that " in any limited, careful use of the term, no candid man could say that the Positive Philosophy contained a great deal as thoroughly antagonistic to [the very essence of[1]] science as Catholicism " ?

M. Comte, it will have been observed, desires to retain the whole of Catholic organization ; and the logical practical result of this part of his doctrine would be the establishment of something corresponding with that eminently Catholic, but admittedly anti-scientific, institution—the Holy Office.

I hope I have said enough to show that I wrote the few lines I devoted to M. Comte and his philosophy, neither unguardedly, nor ignorantly, still less maliciously. I shall be sorry if what I have now added, in my own justification, should lead any to suppose that I think M. Comte's works worthless ; or that I do not heartily respect, and sympathise with, those who have been impelled by him to think deeply upon social problems, and to strive nobly for social regeneration. It is the virtue of that impulse, I believe, which will save the name and fame of Auguste Comte from oblivion. As for his philosophy, I part with it by quoting his own words, reported to me by a quondam Comtist,

---

[1] Mr. Congreve leaves out these important words, which show that I refer to the spirit, and not to the details of science.

now an eminent member of the Institute of France, M. Charles Robin :—

"La Philosophie est une tentative incessante de l'esprit humain pour arriver au repos : mais elle se trouve incessamment aussi dérangée par les progrès continus de la science.   De là vient pour le philosophe l'obligation de refaire chaque soir la synthèse de ses conceptions ; et un jour viendra où l'homme raisonnable ne fera plus d'autre prière du soir."

IX.

ON A PIECE OF CHALK.

A LECTURE TO WORKING MEN.

IF a well were to be sunk at our feet in the midst of the city of Norwich, the diggers would very soon find themselves at work in that white substance almost too soft to be called rock, with which we are all familiar as "chalk."

Not only here, but over the whole county of Norfolk, the well-sinker might carry his shaft down many hundred feet without coming to the end of the chalk; and, on the sea-coast, where the waves have pared away the face of the land which breasts them, the scarped faces of the high cliffs are often wholly formed of the same material. Northward, the chalk may be followed as far as Yorkshire; on the south coast it appears abruptly in the picturesque western bays of Dorset, and breaks into the Needles of the Isle of Wight; while on the shores of Kent it supplies that long line of white cliffs to which England owes her name of Albion.

Were the thin soil which covers it all washed away, a curved band of white chalk, here broader, and there narrower, might be followed diagonally across England from Lulworth in Dorset, to Flamborough Head in Yorkshire—a distance of over 280 miles as the crow flies.

From this band to the North Sea, on the east, and the Channel, on the south, the chalk is largely hidden by other deposits ; but, except in the Weald of Kent and Sussex, it enters into the very foundation of all the south-eastern counties.

Attaining, as it does in some places, a thickness of more than a thousand feet, the English chalk must be admitted to be a mass of considerable magnitude. Nevertheless, it covers but an insignificant portion of the whole area occupied by the chalk formation of the globe, which has precisely the same general characters as ours, and is found in detached patches, some less, and others more extensive, than the English.

Chalk occurs in north-west Ireland ; it stretches over a large part of France,—the chalk which underlies Paris being, in fact, a continuation of that of the London basin ; it runs through Denmark and Central Europe, and extends southward to North Africa ; while, eastward, it appears in the Crimea and in Syria, and may be traced as far as the shores of the Sea of Aral, in Central Asia.

If all the points at which true chalk occurs were circumscribed, they would lie within an irregular oval about 3,000 miles in long diameter—the area of which would be as great as that of Europe, and would many times exceed that of the largest existing inland sea— the Mediterranean.

Thus the chalk is no unimportant element in the masonry of the earth's crust, and it impresses a peculiar stamp, varying with the conditions to which it is exposed, on the scenery of the districts in which it occurs.   The undulating downs and rounded coombs, covered with sweet-grassed turf, of our inland chalk country, have a peacefully domestic and mutton-suggesting prettiness, but can hardly be called either grand or beautiful.   But, on our southern coasts, the wall-sided cliffs, many hundred feet high, with vast needles and pinnacles standing out in the sea, sharp and solitary enough to serve as perches for the wary cormorant, confer a wonderful beauty and grandeur upon the chalk headlands.   And, in the East, chalk has its share in the formation of some of the most venerable of mountain ranges, such as the Lebanon.

What is this wide-spread component of the surface of the earth ? and whence did it come ?

You may think this no very hopeful inquiry.   You may not unnaturally suppose that the attempt to solve such problems as these can lead to no result, save that of entangling the inquirer in vague speculations, incapable of refutation and of verification.

If such were really the case, I should have selected some other subject than a " piece of chalk " for my discourse.   But, in truth, after much deliberation, I have been unable to think of any topic which would so well enable me to lead you to see how solid is the foundation upon which some of the most startling conclusions of physical science rest.

A great chapter of the history of the world is written in the chalk.   Few passages in the history of man can be supported by such an overwhelming mass of direct and indirect evidence as that which testifies to the truth of the fragment of the history of the globe, which I hope to enable you to read, with your own eyes, to-night.

Let me add, that few chapters of human history have a more profound significance for ourselves. I weigh my words well when I assert, that the man who should know the true history of the bit of chalk which every carpenter carries about in his breeches-pocket, though ignorant of all other history, is likely, if he will think his knowledge out to its ultimate results, to have a truer, and therefore a better, conception of this wonderful universe, and of man's relation to it, than the most learned student who is deep-read in the records of humanity and ignorant of those of Nature.

The language of the chalk is not hard to learn, not nearly so hard as Latin, if you only want to get at the broad features of the story it has to tell; and I propose that we now set to work to spell that story out together.

We all know that if we "burn" chalk the result is quicklime. Chalk, in fact, is a compound of carbonic acid gas and lime, and when you make it very hot the carbonic acid flies away and the lime is left.

By this method of procedure we see the lime, but we do not see the carbonic acid. If, on the other hand, you were to powder a little chalk, and drop it into a good deal of strong vinegar, there would be a great bubbling and fizzing, and, finally, a clear liquid, in which no sign of chalk would appear. Here you see the carbonic acid in the bubbles; the lime, dissolved in the vinegar, vanishes from sight. There are a great many other ways of showing that chalk is essentially nothing but carbonic acid and quicklime. Chemists enunciate the result of all the experiments which prove this, by stating that chalk is almost wholly composed of " carbonate of lime."

It is desirable for us to start from the knowledge of this fact, though it may not seem to help us very far towards what we seek. For carbonate of lime is a widely-spread substance, and is met with under very various conditions. All sorts of limestones are composed of more or less pure carbonate of lime. The crust which is often deposited by waters which have drained through limestone rocks, in the form of what are called stalagmites and stalactites, is carbonate of lime. Or, to take a more familiar example, the fur on the inside of a tea-kettle is carbonate of lime; and, for anything chemistry tells us to the contrary, the chalk might be a kind of gigantic fur upon the bottom of the earth-kettle, which is kept pretty hot below.

Let us try another method of making the chalk tell us its own history. To the unassisted eye chalk looks simply like a very loose and open kind of stone. But it is possible to grind a slice of chalk down so thin that you can see through it—until it is thin enough, in fact, to be examined with any magnifying power that may be thought desirable. A thin slice of the fur of a kettle might be made in the same way. If it were examined microscopically, it would show itself to be a more or less distinctly laminated mineral substance, and nothing more.

But the slice of chalk presents a totally different appearance when placed under the microscope. The general mass of it is made up of very minute granules;

but, imbedded in this matrix, are innumerable bodies, some smaller and some larger, but, on a rough average, not more than a hundredth of an inch in diameter, having a well-defined shape and structure. A cubic inch of some specimens of chalk may contain hundreds of thousands of these bodies, compacted together with incalculable millions of the granules.

The examination of a transparent slice gives a good notion of the manner in which the components of the chalk are arranged, and of their relative proportions. But, by rubbing up some chalk with a brush in water and then pouring off the milky fluid, so as to obtain sediments of different degrees of fineness, the granules and the minute rounded bodies may be pretty well separated from one another, and submitted to microscopic examination, either as opaque or as transparent objects. By combining the views obtained in these various methods, each of the rounded bodies may be proved to be a beautifully-constructed calcareous fabric, made up of a number of chambers, communicating freely with one another. The chambered bodies are of various forms. One of the commonest is something like a badly-grown raspberry, being formed of a number of nearly globular chambers of different sizes congregated together. It is called *Globigerina*, and some specimens of chalk consist of little else than *Globigerinæ* and granules.

Let us fix our attention upon the *Globigerina*. It is the spoor of the game we are tracking. If we can learn what it is and what are the conditions of its existence, we shall see our way to the origin and past history of the chalk.

A suggestion which may naturally enough present itself is, that these curious bodies are the result of some process of aggregation which has taken place in the carbonate of lime ; that, just as in winter, the rime on our windows simulates the most delicate and elegantly arborescent foliage—proving that the mere mineral water may, under certain conditions, assume the outward form of organic bodies—so this mineral substance, carbonate of lime, hidden away in the bowels of the earth, has aken the shape of these chambered bodies. I am not raising a merely fanciful and unreal objection. Very learned men, in former days, have even entertained the notion that all the formed things found in rocks are of this nature ; and if no such conception is at present held to be admissible, it is because long and varied experience has now shown that mineral matter never does assume the form and structure we find in fossils. If any one were to try to persuade you that an oyster-shell (which is also chiefly composed of carbonate of lime) had crystallized out of sea-water, I suppose you would laugh at the absurdity. Your laughter would be justified by the fact that all experience tends to show that oyster-shells are formed by the agency of oysters, and in no other way. And if there were no better reasons, we should be justified, on like grounds, in believing that *Globigerina* is not the product of any-thing but vital activity.

Happily, however, better evidence in proof of the organic nature of the *Globigerinæ* than that of analogy is forthcoming. It so happens that calcareous skeletons, exactly similar to the *Globigerinæ* of the chalk, are being formed, at the present moment, by minute living creatures, which flourish in multitudes, literally more numerous than the sands of the sea-shore, over a large extent of that part of the earth's surface which is covered by the ocean.

The history of the discovery of these living *Globigerinœ,* and of the part which they play in rock-building, is singular enough. It is a discovery which, like others of no less scientific importance, has arisen, incidentally, out of work devoted to very different and exceedingly practical interests.

When men first took to the sea, they speedily learned to look out for shoals and rocks; and the more the burthen of their ships increased, the more imperatively necessary it became for sailors to ascertain with precision the depth of the waters they traversed. Out of this necessity grew the use of the lead and sounding-line; and, ultimately, marine-surveying, which is the recording of the form of coasts and of the depth of the sea, as ascertained by the sounding-lead, upon charts.

At the same time, it became desirable to ascertain and to indicate the nature of the sea-bottom, since this circumstance greatly affects its goodness as holding ground for anchors. Some ingenious tar, whose name deserves a better fate than the oblivion into which it has fallen, attained this object by "arming" the bottom of the lead with a lump of grease, to which more or less of the sand or mud, or broken shells, as the case might be, adhered, and was brought to the surface. But, however well adapted such an apparatus might be for rough nautical purposes, scientific accuracy could not be expected from the armed lead, and to remedy its defects (especially when applied to sounding in great depths)

Lieut. Brooke, of the American Navy, some years ago invented a most ingenious machine, by which a considerable portion of the superficial layer of the sea-bottom can be scooped out and brought up, from any depth to which the lead descends.

In 1853, Lieut. Brooke obtained mud from the bottom of the North Atlantic, between Newfoundland and the Azores, at a depth of more than 10,000 feet, or two miles, by the help of this sounding apparatus. The specimens were sent for examination to Ehrenberg of Berlin, and to Bailey of West Point, and those able microscopists found that this deep-sea mud was almost entirely composed of the skeletons of living organisms—the greater proportion of these being just like the *Globigerinæ* already known to occur in the chalk.

Thus far, the work had been carried on simply in the interests of science, but Lieut. Brooke's method of sounding acquired a high commercial·value, when the enterprise of laying down the telegraph-cable between this country and the United States was undertaken. For it became a matter of immense importance to know, not only the depth of the sea over the whole line along which the cable was to be laid, but the exact nature of the bottom, so as to guard against chances of cutting or fraying the strands of that costly rope. The Admiralty consequently ordered Captain Dayman, an old friend and shipmate of mine, to ascertain the depth over the whole line of the cable, and to bring back-specimens of the bottom. In former days, such a command as this might have sounded very much like one of the impossible things which the young prince in the Fairy Tales is ordered to do before he can obtain the hand of the Princess. However, in the months of June and July 1857, my friend performed the task assigned to him with great expedition and precision, without, so far as I know, having met with any reward of that kind. The specimens of Atlantic mud which he procured were sent to me to be examined and reported upon.[1]

The result of all these operations is, that we know the contours and the nature of the surface-soil covered by the North Atlantic, for a distance of 1,700 miles from east to west, as well as we know that of any part of the dry land.

It is a prodigious plain—one of the widest and most even plains in the world. If the sea were drained off, you might drive a wagon all the way from Valentia, on the west coast of Ireland, to Trinity Bay, in Newfoundland. And, except upon one sharp incline about 200 miles from Valentia, I am not quite sure that it would even be necessary to put the skid on, so gentle are the ascents and descents upon that long route. From Valentia the road would lie down hill for about 200 miles to the point at which the bottom is now covered by 1,700 fathoms of sea-water. Then would come the central plain, more than a thousand miles wide, the inequalities of the surface of which would be hardly perceptible, though the depth of water upon it now varies from 10,000 to 15,000 feet; and there are places in which

---

[1] See Appendix to Captain Dayman's " Deep Sea Soundings in the North Atlantic Ocean, between Ireland and Newfoundland, made in H.M.S. *Cyclops.* Published by order of the Lords Commissioners of the Admiralty, 1858." They have since formed the subject of an elaborate Memoir by Messrs. Parker and Jones, published in the *Philosophical Transactions* for 1865.

Mont Blanc might be sunk without showing its peak above water. Beyond this, the ascent on the American side commences, and gradually leads, for about 300 miles, to the Newfoundland shore.

Almost the whole of the bottom of this central plain (which extends for many hundred miles in a north and south direction) is covered by a fine mud, which, when brought to the surface, dries into a greyish-white friable substance. You can write with this on a blackboard, if you are so inclined ; and, to the eye, it is quite like very soft, greyish chalk. Examined chemically, it proves to be composed almost wholly of carbonate of lime ; and if you make a section of it, in the same way as that of the piece of chalk was made, and view it with the microscope, it presents innumerable *Globigerinæ*, embedded in a granular matrix.

Thus this deep-sea mud is substantially chalk. I say substantially, because there are a good many minor differences : but as these have no bearing on the question immediately before us,—which is the nature of the *Globigerinæ* of the chalk,—it is unnecessary to speak of them.

Globigerinæ of every size, from the smallest to the largest, are associated together in the Atlantic mud, and the chambers of many are filled by a soft animal matter. This soft substance is, in fact, the remains of the creature to which the *Globigerina* shell, or rather skeleton, owes its existence—and which is an animal of the simplest imaginable description. It is, in fact, a mere particle of living jelly, without defined parts of any kind— without a mouth, nerves, muscles, or distinct organs, and only manifesting its vitality to ordinary observation by thrusting out and retracting from all parts of its surface, long filamentous processes, which serve for arms and legs. Yet this amorphous particle, devoid of everything which, in the higher animals, we call organs, is capable of feeding, growing, and multiplying; of separating from the ocean the small proportion of carbonate of lime which is dissolved in sea-water; and of building up that substance into a skeleton for itself, according to a pattern which can be imitated by no other known agency.

The notion that animals can live and flourish in the sea, at the vast depths from which apparently living *Globigerinæ* have been brought up, does not agree very well with our usual conceptions respecting the conditions of animal life; and it is not so absolutely impossible as it might at first sight appear to be, that the *Globigerinæ* of the Atlantic sea-bottom do not live and die where they are found.

As I have mentioned, the soundings from the great Atlantic plain are almost entirely made up of *Globigerinæ*, with the granules which have been mentioned, and some few other calcareous shells; but a small percentage of the chalky mud—perhaps at most some five per cent. of it—is of a different nature, and consists of shells and skeletons composed of silex, or pure flint. These silicious bodies belong partly to the lowly vegetable organisms which are called *Diatomaceæ*, and partly to the minute, and extremely simple, animals, termed *Radiolaria*. It is quite certain that these creatures do not live at the bottom of the ocean, but at its surface—where they may be obtained in prodigious numbers by the use of a properly constructed net.

Hence it follows that these silicious organisms, though they are not heavier than the lightest dust, must have fallen, in some cases, through fifteen thousand feet of water, before they reached their final resting-place on the ocean floor. And, considering how large a surface these bodies expose in proportion to their weight, it is probable that they occupy a great length of time in making their burial journey from the surface of the Atlantic to the bottom.

But if the *Radiolaria* and Diatoms are thus rained upon the bottom of the sea, from the superficial layer of its waters in which they pass their lives, it is obviously possible that the *Globigerinæ* may be similarly derived ; and if they were so, it would be much more easy to understand how they obtain their supply of food than it is at present. Nevertheless, the positive and negative evidence all points the other way. The skeletons of the full-grown, deep-sea *Globigerinæ* are so remarkably solid and heavy in proportion to their surface as to seem little fitted for floating ; and, as a matter of fact, they are not to be found along with the Diatoms and *Radiolaria*, in the uppermost stratum of the open ocean.

It has been observed, again, that the abundance of *Globigerinæ*, in proportion to other organisms of like kind, increases with the depth of the sea ; and that deep-water *Globigerinæ* are larger than those which live in shallower parts of the sea ; and such facts negative the supposition that these organisms have been swept by currents from the shallows into the deeps of the Atlantic.

It therefore seems to be hardly doubtful that these

wonderful creatures live and die at the depths in which
they are found.[1]

However, the important points for us are, that the
living *Globigerinæ* are exclusively marine animals, the
skeletons of which abound at the bottom of deep seas;
and that there is not a shadow of reason for believing
that the habits of the *Globigerinæ* of the chalk differed
from those of the existing species. But if this be true,
there is no escaping the conclusion that the chalk itself
is the dried mud of an ancient deep sea.

In working over the soundings collected by Captain
Dayman, I was surprised to find that many of what
I have called the "granules" of that mud, were not, as
one might have been tempted to think at first, the mere
powder and waste of *Globigerinæ*, but that they had a
definite form and size. I termed these bodies "*coccoliths*," and doubted their organic nature. Dr. Wallich
verified my observation, and added the interesting
discovery that, not unfrequently, bodies similar to these
"coccoliths" were aggregated together into spheroids,
which he termed "*coccospheres.*" So far as we knew,
these bodies, the nature of which is extremely puzzling
and problematical, were peculiar to the Atlantic
soundings.

[1] During the cruise of H.M.S. *Bull-dog*, commanded by Sir Leopold
M'Clintock, in 1860, living star-fish were brought up, clinging to the lowest
part of the sounding-line, from a depth of 1,260 fathoms, midway between
Cape Farewell, in Greenland, and the Rockall banks. Dr. Wallich ascertained
that the sea-bottom at this point consisted of the ordinary *Globigerina* ooze,
and that the stomachs of the star-fishes were full of *Globigerinæ*. This
discovery removes all objections to the existence of living *Globigerinæ* at
great depths, which are based upon the supposed difficulty of maintaining
animal life under such conditions; and it throws the burden of proof upon
those who object to the supposition that the *Globigerinæ* live and die where
they are found.

But, a few years ago, Mr. Sorby, in making a careful examination of the chalk by means of thin sections and otherwise, observed, as Ehrenberg had done before him, that much of its granular basis possesses a definite form. Comparing these formed particles with those in the Atlantic soundings, he found the two to be identical; and thus proved that the chalk, like the soundings, contains these mysterious coccoliths and coccospheres. Here was a further and a most interesting confirmation, from internal evidence, of the essential identity of the chalk with modern deep-sea mud. *Globigerinæ*, coccoliths, and coccospheres are found as the chief constituents of both, and testify to the general similarity of the conditions under which both have been formed.[1]

The evidence furnished by the hewing, facing, and superposition of the stones of the Pyramids, that these structures were built by men, has no greater weight than the evidence that the chalk was built by *Globigerinæ*; and the belief that those ancient pyramid-builders were terrestrial and air-breathing creatures like ourselves, is not better based than the conviction that the chalk-makers lived in the sea.

But as our belief in the building of the Pyramids by men is not only grounded on the internal evidence afforded by these structures, but gathers strength from multitudinous collateral proofs, and is clinched by the total absence of any reason for a contrary belief; so the evidence drawn from the *Globeriginæ* that the chalk is an ancient sea-bottom, is fortified by innumerable inde-

---

[1] I have recently traced out the development of the "coccoliths" from a diameter of $\frac{1}{7000}$th of an inch up to their largest size (which is about $\frac{1}{1600}$th), and no longer doubt that they are produced by independent organisms, which, like the *Globigerinæ*, live and die at the bottom of the sea.

pendent lines of evidence ; and our belief in the truth
of the conclusion to which all positive testimony tends,
receives the like negative justification from the fact that
no other hypothesis has a shadow of foundation.

It may be worth while briefly to consider a few of
these collateral proofs that the chalk was deposited at
the bottom of the sea.

The great mass of the chalk is composed, as we have
seen, of the skeletons of *Globigerinæ*, and other simple
organisms, imbedded in granular matter.   Here and
there, however, this hardened mud of the ancient sea
reveals the remains of higher animals which have lived
and died, and left their hard parts in the mud, just as
the oysters die and leave their shells behind them, in the
mud of the present seas.

There are, at the present day, certain groups of animals
which are never found in fresh waters, being unable to
live anywhere but in the sea.   Such are the corals ; those
corallines which are called *Polyzoa ;* those creatures
which fabricate the lamp-shells, and are called *Brachio-
poda ;* the pearly *Nautilus*, and all animals allied to
it ; and all the forms of sea-urchins and star-fishes.

Not only are all these creatures confined to salt water
at the present day ; but, so far as our records of the past
go, the conditions of their existence have been the same :
hence, their occurrence in any deposit is as strong
evidence as can be obtained, that that deposit was
formed in the sea.   Now the remains of animals of all
the kinds which have been enumerated, occur in the
chalk, in greater or less abundance ; while not one of
those forms of shell-fish which are characteristic of fresh
water has yet been observed in it.

When we consider that the remains of more than three thousand distinct species of aquatic animals have been discovered among the fossils of the chalk, that the great majority of them are of such forms as are now met with only in the sea, and that there is no reason to believe that any one of them inhabited fresh water—the collateral evidence that the chalk represents an ancient sea-bottom acquires as great force as the proof derived from the nature of the chalk itself. I think you will now allow that I did not overstate my case when I asserted that we have as strong grounds for believing that all the vast area of dry land, at present occupied by the chalk, was once at the bottom of the sea, as we have for any matter of history whatever ; while there is no justification for any other belief.

No less certain is it that the time during which the countries we now call south-east England, France, Germany, Poland, Russia, Egypt, Arabia, Syria, were more or less completely covered by a deep sea, was of considerable duration.

We have already seen that the chalk is, in places, more than a thousand feet thick. I think you will agree with me, that it must have taken some time for the skeletons of animalcules of a hundredth of an inch in diameter to heap up such a mass as that. I have said that throughout the thickness of the chalk the remains of other animals are scattered. These remains are often in the most exquisite state of preservation. The valves of the shell-fishes are commonly adherent ; the long spines of some of the sea-urchins, which would be detached by the smallest jar, often remain in their places. In a word, it is certain that these animals have lived

and died when the place which they now occupy was
the surface of as much of the chalk as had then been
deposited ; and that each has been covered up by the
layer of *Globigerina* mud, upon which the creatures
imbedded a little higher up have, in like manner, lived
and died. But some of these remains prove the existence
of reptiles of vast size in the chalk sea. These lived
their time, and had their ancestors and descendants,
which assuredly implies time, reptiles being of slow
growth.

There is more curious evidence, again, that the process
of covering up, or, in other words, the deposit of *Globigerina* skeletons, did not go on very fast. It is demonstrable that an animal of the cretaceous sea might die,
that its skeleton might lie uncovered upon the sea-bottom
long enough to lose all its outward coverings and appendages by putrefaction ; and that, after this had happened,
another animal might attach itself to the dead and naked
skeleton, might grow to maturity, and might itself die
before the calcareous mud had buried the whole.

Cases of this kind are admirably described by Sir
Charles Lyell. He speaks of the frequency with which
geologists find in the chalk a fossilized sea-urchin, to
which is attached the lower valve of a *Crania*. This
is a kind of shell-fish, with a shell composed of two
pieces, of which, as in the oyster, one is fixed and the
other free.

"The upper valve is almost invariably wanting,
though occasionally found in a perfect state of preservation in the white chalk at some distance. In this
case, we see clearly that the sea-urchin first lived from
youth to age, then died and lost its spines, which were

carried away. Then the young *Crania* adhered to the bared shell, grew and perished in its turn; after which, the upper valve was separated from the lower, before the Echinus became enveloped in chalky mud."[1]

A specimen in the Museum of Practical Geology, in London, still further prolongs the period which must have elapsed between the death of the sea-urchin, and its burial by the *Globigerinæ*. For the outward face of the valve of a *Crania*, which is attached to a sea-urchin (*Micraster*), is itself overrun by an incrusting coralline, which spreads thence over more or less of the surface of the sea-urchin. It follows that, after the upper valve of the *Crania* fell off, the surface of the attached valve must have remained exposed long enough to allow of the growth of the whole coralline, since corallines do not live imbedded in mud.

The progress of knowledge may, one day, enable us to deduce from such facts as these the maximum rate at which the chalk can have accumulated, and thus to arrive at the minimum duration of the chalk period. Suppose that the valve of the *Crania* upon which a coralline has fixed itself in the way just described, is so attached to the sea-urchin that no part of it is more than an inch above the face upon which the sea-urchin rests. Then, as the coralline could not have fixed itself, if the *Crania* had been covered up with chalk mud, and could not have lived had itself been so covered, it follows, that an inch of chalk mud could not have accumulated within the time between the death and decay of the soft parts of the sea-urchin and the growth of the coralline to the full size which it has attained. If the decay of the soft parts of the sea-urchin; the attachment, growth to

[1] " Elements of Geology," by Sir Charles Lyell, Bart. F.R.S., p. 23.

maturity, and decay of the *Crania*; and the subsequent attachment and growth of the coralline, took a year (which is a low estimate enough), the accumulation of the inch of chalk must have taken more than a year : and the deposit of a thousand feet of chalk must, consequently, have taken more than twelve thousand years.

The foundation of all this calculation is, of course, a knowledge of the length of time the *Crania* and the coralline needed to attain their full size ; and, on this head, precise knowledge is at present wanting. But there are circumstances which tend to show, that nothing like an inch of chalk has accumulated during the life of a *Crania*; and, on any probable estimate of the length of that life, the chalk period must have had a much longer duration than that thus roughly assigned to it.

Thus, not only is it certain that the chalk is the mud of an ancient sea-bottom ; but it is no less certain, that the chalk sea existed during an extremely long period, though we may not be prepared to give a precise estimate of the length of that period in years. The relative duration is clear, though the absolute duration may not be definable. The attempt to affix any precise date to the period at which the chalk sea began, or ended, its existence, is baffled by difficulties of the same kind. But the relative age of the cretaceous epoch may be determined with as great ease and certainty as the long duration of that epoch.

You will have heard of the interesting discoveries recently made, in various parts of Western Europe, of flint implements, obviously worked into shape by human hands, under circumstances which show conclusively that man is a very ancient denizen of these regions.

It has been proved that the old populations of Europe, whose existence has been revealed to us in this way, consisted of savages, such as the Esquimaux are now; that, in the country which is now France, they hunted the reindeer, and were familiar with the ways of the mammoth and the bison. The physical geography of France was in those days different from what it is now—the river Somme, for instance, having cut its bed a hundred feet deeper between that time and this; and, it is probable, that the climate was more like that of Canada or Siberia, than that of Western Europe.

The existence of these people is forgotten even in the traditions of the oldest historical nations. The name and fame of them had utterly vanished until a few years back; and the amount of physical change which has been effected since their day, renders it more than probable that, venerable as are some of the historical nations, the workers of the chipped flints of Hoxne or of Amiens are to them, as they are to us, in point of antiquity.

But, if we assign to these hoar relics of long vanished generations of men the greatest age that can possibly be claimed for them, they are not older than the drift, or boulder clay, which, in comparison with the chalk, is but a very juvenile deposit. You need go no further than your own sea-board for evidence of this fact. At one of the most charming spots on the coast of Norfolk, Cromer, you will see the boulder clay forming a vast mass, which lies upon the chalk, and must consequently have come into existence after it. Huge boulders of chalk are, in fact, included in the clay, and have evidently been brought to the position they now occupy, by the same agency as that which has planted blocks of syenite from Norway side by side with them.

The chalk, then, is certainly older than the boulder clay. If you ask how much, I will again take you no further than the same spot upon your own coasts for evidence. I have spoken of the boulder clay and drift as resting upon the chalk. That is not strictly true. Interposed between the chalk and the drift is a comparatively insignificant layer, containing vegetable matter. But that layer tells a wonderful history. It is full of stumps of trees standing as they grew. Fir-trees are there with their cones, and hazel-bushes with their nuts ; there stand the stools of oak and yew trees, beeches and alders. Hence this stratum is appropriately called the " forest-bed."

It is obvious that the chalk must have been upheaved and converted into dry land, before the timber trees could grow upon it. As the bolls of some of these trees are from two to three feet in diameter, it is no less clear that the dry land thus formed remained in the same condition for long ages. And not only do the remains of stately oaks and well-grown firs testify to the duration of this condition of things, but additional evidence to the same effect is afforded by the abundant remains of elephants, rhinoceroses, hippopotamuses, and other great wild beasts, which it has yielded to the zealous search of such men as the Rev. Mr. Gunn.

When you look at such a collection as he has formed, and bethink you that these elephantine bones did veritably carry their owners about, and these great grinders crunch, in the dark woods of which the forest-bed is now the only trace, it is impossible not to feel that they are as good evidence of the lapse of time as the annual rings of the tree-stumps.

Thus there is a writing upon the wall of cliffs at Cromer, and whoso runs may read it. It tells us, with an authority which cannot be impeached, that the ancient sea-bed of the chalk sea was raised up, and remained dry land, until it was covered with forest, stocked with the great game whose spoils have rejoiced your geologists. How long it remained in that condition cannot be said ; but " the whirligig of time brought its revenges " in those days as in these. That dry land, with the bones and teeth of generations of long-lived elephants, hidden away among the gnarled roots and dry leaves of its ancient trees, sank gradually to the bottom of the icy sea, which covered it with huge masses of drift and boulder clay. Sea-beasts, such as the walrus, now restricted to the extreme north, paddled about where birds had twittered among the topmost twigs of the fir-trees. How long this state of things endured we know not, but at length it came to an end. The upheaved glacial mud hardened into the soil of modern Norfolk. Forests grew once more, the wolf and the beaver re-placed the reindeer and the elephant ; and at length what we call the history of England dawned.

Thus you have, within the limits of your own county, proof that the chalk can justly claim a very much greater antiquity than even the oldest physical traces of mankind. But we may go further and demonstrate, by evidence of the same authority as that which testifies to the existence of the father of men, that the chalk is vastly older than Adam himself.

The Book of Genesis informs us that Adam, immediately upon his creation, and before the appearance of Eve, was placed in the Garden of Eden. The problem of the geographical position of Eden has greatly vexed the spirits of the learned in such matters, but there is one point respecting which, so far as I know, no commentator has ever raised a doubt. This is, that of the four rivers which are said to run out of it, Euphrates and Hiddekel are identical with the rivers now known by the names of Euphrates and Tigris.

But the whole country in which these mighty rivers take their origin, and through which they run, is composed of rocks which are either of the same age as the chalk, or of later date. So that the chalk must not only have been formed, but, after its formation, the time required for the deposit of these later rocks, and for their upheaval into dry land, must have elapsed, before the smallest brook which feeds the swift stream of "the great river, the river of Babylon," began to flow.

Thus, evidence which cannot be rebutted, and which need not be strengthened, though if time permitted I might indefinitely increase its quantity, compels you to believe that the earth, from the time of the chalk to the present day, has been the theatre of a series of changes as vast in their amount, as they were slow in their progress. The area on which we stand has been first sea and then land, for at least four alternations; and has remained in each of these conditions for a period of great length.

Nor have these wonderful metamorphoses of sea into land, and of land into sea, been confined to one corner of England. During the chalk period, or "cretaceous epoch," not one of the present great physical features of the globe was in existence. Our great mountain ranges, Pyrenees, Alps, Himalayas, Andes, have all been up-heaved since the chalk was deposited, and the cretaceous sea flowed over the sites of Sinai and Ararat.

All this is certain, because rocks of cretaceous, or still later, date have shared in the elevatory movements which gave rise to these mountain chains; and may be found perched up, in some cases, many thousand feet high upon their flanks. And evidence of equal cogency demonstrates that, though, in Norfolk, the forest-bed rests directly upon the chalk, yet it does so, not because the period at which the forest grew imme-diately followed that at which the chalk was formed, but because an immense lapse of time, represented elsewhere by thousands of feet of rock, is not indicated at Cromer.

I must ask you to believe that there is no less con-clusive proof that a still more prolonged succession of similar changes occurred, before the chalk was deposited. Nor have we any reason to think that the first term in the series of these changes is known. The oldest sea-beds preserved to us are sands, and mud, and pebbles, the wear and tear of rocks which were formed in still older oceans.

But, great as is the magnitude of these physical changes of the world, they have been accompanied by a no less striking series of modifications in its living inhabitants.

All the great classes of animals, beasts of the field, fowls of the air, creeping things, and things which dwell in the waters, flourished upon the globe long ages before the chalk was deposited. Very few, however, if any, of these ancient forms of animal life were identical with those which now live. Certainly not one of the higher animals was of the same species as any of those now in existence. The beasts of the field, in the days before the chalk, were not our beasts of the field, nor the fowls of the air such as those which the eye of men has seen flying, unless his antiquity dates infinitely further back than we at present surmise. If we could be carried back into those times, we should be as one suddenly set down in Australia before it was colonized. We should see mammals, birds, reptiles, fishes, insects, snails, and the like, clearly recognisable as such, and yet not one of them would be just the same as those with which we are familiar, and many would be extremely different.

From that time to the present, the population of the world has undergone slow and gradual, but incessant, changes. There has been no grand catastrophe—no destroyer has swept away the forms of life of one period, and replaced them by a totally new creation; but one species has vanished and another has taken its place; creatures of one type of structure have diminished, those of another have increased, as time has passed on. And thus, while the differences between the living creatures of the time before the chalk and those of the present day appear startling, if placed side by side, we are led from one to the other by the most gradual progress, if we follow the course of Nature through the whole series of those relics of her operations which she has left behind.

And it is by the population of the chalk sea that the ancient and the modern inhabitants of the world are most completely connected. The groups which are dying out flourish, side by side, with the groups which are now the dominant forms of life.

Thus the chalk contains remains of those strange flying and swimming reptiles, the pterodactyl, the ichthyosaurus, and the plesiosaurus, which are found in no later deposits, but abounded in preceding ages. The chambered shells called ammonites and belemnites, which are so characteristic of the period preceding the cretaceous, in like manner die with it.

But, amongst these fading remainders of a previous state of things, are some very modern forms of life, looking like Yankee pedlars among a tribe of Red Indians. Crocodiles of modern type appear; bony fishes, many of them very similar to existing species, almost supplant the forms of fish which predominate in more ancient seas; and many kinds of living shell-fish first become known to us in the chalk. The vegetation acquires a modern aspect. A few living animals are not even distinguishable as species, from those which existed at that remote epoch. The *Globigerina* of the present day, for example, is not different specifically from that of the chalk; and the same may be said of many other *Foraminifera*. I think it probable that critical and unprejudiced examination will show that more than one species of much higher animals have had a similar longevity; but the only example which I can at present give confidently is the snake's-head lamp-shell (*Terebratulina caput serpentis*), which lives in our English seas and abounded (as *Terebratulina striata* of authors) in the chalk.

The longest line of human ancestry must hide its diminished head before the pedigree of this insignificant shell-fish. We Englishmen are proud to have an ancestor who was present at the Battle of Hastings. The ancestors of *Terebratulina caput serpentis* may have been present at a battle of *Ichthyosauria* in that part of the sea which, when the chalk was forming, flowed over the site of Hastings. While all around has changed, this *Terebratulina* has peacefully propagated its species from generation to generation, and stands to this day, as a living testimony to the continuity of the present with the past history of the globe.

Up to this moment I have stated, so far as I know, nothing but well-authenticated facts, and the immediate conclusions which they force upon the mind.

But the mind is so constituted that it does not willingly rest in facts and immediate causes, but seeks always after a knowledge of the remoter links in the chain of causation.

Taking the many changes of any given spot of the earth's surface, from sea to land and from land to sea, as an established fact, we cannot refrain from asking ourselves how these changes have occurred. And when we have explained them—as they must be explained —by the alternate slow movements of elevation and depression which have affected the crust of the earth, we go still further back, and ask, Why these movements?

I am not certain that any one can give you a satisfactory answer to that question. Assuredly I cannot. All that can be said, for certain, is, that such movements are part of the ordinary course of nature, inasmuch as they are going on at the present time. Direct proof may be given, that some parts of the land of the northern hemisphere are at this moment insensibly rising and others insensibly sinking; and there is indirect, but perfectly satisfactory, proof, that an enormous area now covered by the Pacific has been deepened thousands of feet, since the present inhabitants of that sea came into existence.

Thus there is not a shadow of a reason for believing that the physical changes of the globe, in past times, have been effected by other than natural causes.

Is there any more reason for believing that the concomitant modifications in the forms of the living inhabitants of the globe have been brought about in other ways?

Before attempting to answer this question, let us try to form a distinct mental picture of what has happened in some special case.

The crocodiles are animals which, as a group, have a very vast antiquity. They abounded ages before the chalk was deposited; they throng the rivers in warm climates, at the present day. There is a difference in the form of the joints of the back-bone, and in some minor particulars, between the crocodiles of the present epoch and those which lived before the chalk; but, in the cretaceous epoch, as I have already mentioned, the crocodiles had assumed the modern type of structure. Notwithstanding this, the crocodiles of the chalk are not identically the same as those which lived in the times called " older tertiary," which succeeded the cretaceous epoch ; and the crocodiles of the older tertiaries are not identical with those of the newer tertiaries, nor are these identical with existing forms. I leave open the question whether particular species may have lived on from epoch to epoch. But each epoch has had its peculiar crocodiles ; though all, since the chalk, have belonged to the modern type, and differ simply in their proportions, and in such structural particulars as are discernible only to trained eyes.

How is the existence of this long succession of different species of crocodiles to be accounted for ?

Only two suppositions seem to be open to us—Either each species of crocodile has been specially created, or it has arisen out of some pre-existing form by the operation of natural causes.

Choose your hypothesis ; I have chosen mine. I can find no warranty for believing in the distinct creation of a score of successive species of crocodiles in the course of countless ages of time. Science gives no countenance to such a wild fancy ; nor can even the perverse ingenuity of a commentator pretend to discover this sense, in the simple words in which the writer of Genesis records the proceedings of the fifth and sixth days of the Creation.

On the other hand, I see no good reason for doubting the necessary alternative, that all these varied species have been evolved from pre-existing crocodilian forms, by the operation of causes as completely a part of the common order of nature, as those which have effected the changes of the inorganic world.

Few will venture to affirm that the reasoning which applies to crocodiles loses its force among other animals, or among plants.   If one series of species has come into existence by the operation of natural causes, it seems folly to deny that all may have arisen in the same way.

A small beginning has led us to a great ending.   If I were to put the bit of chalk with which we started into the hot but obscure flame of burning hydrogen, it would presently shine like the sun.   It seems to me that this physical metamorphosis is no false image of what has been the result of our subjecting it to a jet of fervent, though nowise brilliant, thought to-night.   It has become luminous, and its clear rays, penetrating the abyss of the remote past, have brought within our ken some stages of the evolution of the earth.   And in the shifting " without haste, but without rest " of the land and sea, as in the endless variation of the forms assumed by living beings, we have observed nothing but the natural product of the forces originally possessed by the substance of the universe.

X.

GEOLOGICAL CONTEMPORANEITY AND PERSISTENT TYPES OF LIFE.

MERCHANTS occasionally go through a wholesome, though troublesome and not always satisfactory, process which they term "taking stock." After all the excitement of speculation, the pleasure of gain, and the pain of loss, the trader makes up his mind to face facts and to learn the exact quantity and quality of his solid and reliable possessions.

The man of science does well sometimes to imitate this procedure; and, forgetting for the time the importance of his own small winnings, to re-examine the common stock in trade, so that he may make sure how far the stock of bullion in the cellar—on the faith of whose existence so much paper has been circulating—is really the solid gold of truth.

The Anniversary Meeting of the Geological Society seems to be an occasion well suited for an undertaking of this kind—for an inquiry, in fact, into the nature and value of the present results of palæontological investigation; and the more so, as all those who have paid close attention to the late multitudinous discussions in which palæontology is implicated, must have felt the urgent necessity of some such scrutiny.

First in order, as the most definite and unquestionable of all the results of palæontology, must be mentioned the immense extension and impulse given to botany, zoology, and comparative anatomy, by the investigation of fossil remains. Indeed, the mass of biological facts has been so greatly increased, and the range of biological speculation has been so vastly widened, by the researches of the geologist and palæontologist, that it is to be feared there are naturalists in existence who look upon geology as Brindley regarded rivers. " Rivers," said the great engineer, " were made to feed canals ; " and geology, some seem to think, was solely created to advance comparative anatomy.

Were such a thought justifiable, it could hardly expect to be received with favour by this assembly. But it is not justifiable. Your favourite science has her own great aims independent of all others ; and if, notwithstanding her steady devotion to her own progress, she can scatter such rich alms among her sisters, it should be remembered that her charity is of the sort that does not impoverish, but " blesseth him that gives and him that takes."

Regard the matter as we will, however, the facts remain. Nearly 40,000 species of animals and plants have been added to the Systema Naturæ by palæontological research. This is a living population equivalent to that of a new continent in mere number ; equivalent to that of a new hemisphere, if we take into account the small population of insects as yet found fossil, and the large proportion and peculiar organization of many of the Vertebrata.

But, beyond this, it is perhaps not too much to say that, except for the necessity of interpreting palæontological facts, the laws of distribution would have received less careful study; while few comparative anatomists (and those not of the first order) would have been induced by mere love of detail, as such, to study the minutiæ of osteology, were it not that in such minutiæ lie the only keys to the most interesting riddles offered by the extinct animal world.

These assuredly are great and solid gains. Surely it is matter for no small congratulation that in half a century (for palæontology, though it dawned earlier, came into full day only with Cuvier) a subordinate branch of biology should have doubled the value and the interest of the whole group of sciences to which it belongs.

But this is not all. Allied with geology, palæontology has established two laws of inestimable importance : the first, that one and the same area of the earth's surface has been successively occupied by very different kinds of living beings ; the second, that the order of succession established in one locality holds good, approximately, in all.

The first of these laws is universal and irreversible ; the second is an induction from a vast number of observations, though it may possibly, and even probably, have to admit of exceptions. As a consequence of the second law, it follows that a peculiar relation frequently subsists between series of strata, containing organic remains, in different localities. The series resemble one another, not only in virtue of a general resemblance of the organic remains in the two, but also in virtue of a resemblance in the order and character of the serial succession in each. There is a resemblance of arrangement; so that the separate terms of each series, as well as the whole series, exhibit a correspondence.

Succession implies time; the lower members of a series of sedimentary rocks are certainly older than the upper; and when the notion of age was once introduced as the equivalent of succession, it was no wonder that correspondence in succession came to be looked upon as correspondence in age, or "contemporaneity." And, indeed, so long as relative age only is spoken of, correspondence in succession *is* correspondence in age; it is *relative* contemporaneity.

But it would have been very much better for geology if so loose and ambiguous a word as "contemporaneous" had been excluded from her terminology, and if, in its stead, some term expressing similarity of serial relation, and excluding the notion of time altogether, had been employed to denote correspondence in position in two or more series of strata.

In anatomy, where such correspondence of position has constantly to be spoken of, it is denoted by the word "homology" and its derivatives; and for Geology (which after all is only the anatomy and physiology of the earth) it might be well to invent some single word, such as "homotaxis" (similarity of order), in order to express an essentially similar idea. This, however, has not been done, and most probably the inquiry will at once be made—To what end burden science with a new and strange term in place of one old, familiar, and part of our common language?

The reply to this question will become obvious as the inquiry into the results of palæontology is pushed further.

Those whose business it is to acquaint themselves specially with the works of palæontologists, in fact, will be fully aware that very few, if any, would rest satisfied with such a statement of the conclusions of their branch of biology as that which has just been given.

Our standard repertories of palæontology profess to teach us far higher things—to disclose the entire succession of living forms upon the surface of the globe; to tell us of a wholly different distribution of climatic conditions in ancient times; to reveal the character of the first of all living existences; and to trace out the law of progress from them to us.

It may not be unprofitable to bestow on these professions a somewhat more critical examination than they have hitherto received, in order to ascertain how far they rest on an irrefragable basis; or whether, after all, it might not be well for palæontologists to learn a little more carefully that scientific "ars artium," the art of saying "I don't know." And to this end let us define somewhat more exactly the extent of these pretensions of palæontology.

Every one is aware that Professor Bronn's "Untersuchungen" and Professor Pictet's "Traité de Paléontologie" are works of standard authority, familiarly consulted by every working palæontologist. It is desirable to speak of these excellent books, and of their

distinguished authors, with the utmost respect, and in
a tone as far as possible removed from carping criticism ;
indeed, if they are specially cited in this place, it is
merely in justification of the assertion that the following propositions, which may be found implicitly, or
explicitly, in the works in question, are regarded by
the mass of palæontologists and geologists, not only
on the Continent but in this country, as expressing
some of the best-established results of palæontology.
Thus :—

Animals and plants began their existence together,
not long after the commencement of the deposition of
the sedimentary rocks ; and then succeeded one another,
in such a manner, that totally distinct faunæ and floræ
occupied the whole surface of the earth, one after the
other, and during distinct epochs of time.

A geological formation is the sum of all the strata
deposited over the whole surface of the earth during
one of these epochs : a geological fauna or flora is the
sum of all the species of animals or plants which
occupied the whole surface of the globe, during one
of these epochs.

The population of the earth's surface was at first
very similar in all parts, and only from the middle of
the Tertiary epoch onwards, began to show a distinct
distribution in zones.

The constitution of the original population, as well
as the numerical proportions of its members, indicates
a warmer and, on the whole, somewhat tropical climate,
which remained tolerably equable throughout the year.
The subsequent distribution of living beings in zones
is the result of a gradual lowering of the general

temperature, which first began to be felt at the poles.

It is not now proposed to inquire whether these doctrines are true or false; but to direct your attention to a much simpler though very essential preliminary question—What is their logical basis? what are the fundamental assumptions upon which they all logically depend? and what is the evidence on which those fundamental propositions demand our assent?

These assumptions are two: the first, that the commencement of the geological record is coeval with the commencement of life on the globe; the second, that geological contemporaneity is the same thing as chronological synchrony. Without the first of these assumptions there would of course be no ground for any statement respecting the commencement of life; without the second, all the other statements cited, every one of which implies a knowledge of the state of different parts of the earth at one and the same time, will be no less devoid of demonstration.

The first assumption obviously rests entirely on negative evidence. This is, of course, the only evidence that ever can be available to prove the commencement of any series of phænomena; but, at the same time, it must be recollected that the value of negative evidence depends entirely on the amount of positive corroboration it receives. If A. B. wishes to prove an *alibi*, it is of no use for him to get a thousand witnesses simply to swear that they did not see him in such and such a place, unless the witnesses are prepared to prove that they must have seen him had he been there. But the evidence that animal life commenced with the Lingula-flags, *e. g.*, would seem to be exactly of this unsatisfactory uncorroborated sort. The Cambrian witnesses simply swear they "haven't seen anybody their way;" upon which the counsel for the other side immediately puts in ten or twelve thousand feet of Devonian sandstones to make oath they never saw a fish or a mollusk, though all the world knows there were plenty in their time.

But then it is urged that, though the Devonian rocks in one part of the world exhibit no fossils, in another they do, while the lower Cambrian rocks nowhere exhibit fossils, and hence no living being could have existed in their epoch.

To this there are two replies : the first, that the observational basis of the assertion that the lowest rocks are nowhere fossiliferous is an amazingly small one, seeing how very small an area, in comparison to that of the whole world, has yet been fully searched; the second, that the argument is good for nothing unless the unfossiliferous rocks in question were not only *contemporaneous* in the geological sense, but *synchronous* in the chronological sense. To use the *alibi* illustration again. If a man wishes to prove he was in neither of two places, A and B, on a given day, his witnesses for each place must be prepared to answer for the whole day. If they can only prove that he was not at A in the morning, and not at B in the afternoon, the evidence of his absence from both is *nil*, because he might have been at B in the morning and at A in the afternoon.

Thus everything depends upon the validity of the

second assumption. And we must proceed to inquire what is the real meaning of the word "contemporaneous" as employed by geologists. To this end a concrete example may be taken.

The Lias of England and the Lias of Germany, the Cretaceous rocks of Britain and the Cretaceous rocks of Southern India, are termed by geologists "contemporaneous" formations; but whenever any thoughtful geologist is asked whether he means to say that they were deposited synchronously, he says, "No,—only within the same great epoch." And if, in pursuing the inquiry, he is asked what may be the approximate value in time of a "great epoch"—whether it means a hundred years, or a thousand, or a million, or ten million years—his reply is, "I cannot tell."

If the further question be put, whether physical geology is in possession of any method by which the actual synchrony (or the reverse) of any two distant deposits can be ascertained, no such method can be heard of; it being admitted by all the best authorities that neither similarity of mineral composition, nor of physical character, nor even direct continuity of stratum, are *absolute* proofs of the synchronism of even approximated sedimentary strata : while, for distant deposits, there seems to be no kind of physical evidence attainable of a nature competent to decide whether such deposits were formed simultaneously, or whether they possess any given difference of antiquity. To return to an example already given. All competent authorities will probably assent to the proposition that physical geology does not enable us in any way to reply to this question—Were the British Cretaceous rocks depo-

sited at the same time as those of India, or are they a
million of years younger or a million of years older?

Is palæontology able to succeed where physical
geology fails? Standard writers on palæontology, as
has been seen, assume that she can. They take it for
granted, that deposits containing similar organic remains
are synchronous—at any rate in a broad sense; and
yet, those who will study the eleventh and twelfth
chapters of Sir Henry De la Beche's remarkable " Researches in Theoretical Geology," published now nearly
thirty years ago, and will carry out the arguments
there most luminously stated, to their logical consequences, may very easily convince themselves that
even absolute identity of organic contents is no proof
of the synchrony of deposits, while absolute diversity
is no proof of difference of date. Sir Henry De la
Beche goes even further, and adduces conclusive evidence
to show that the different parts of one and the same
stratum, having a similar composition throughout, containing the same organic remains, and having similar
beds above and below it, may yet differ to any conceivable extent in age.

Edward Forbes was in the habit of asserting that
the similarity of the organic contents of distant formations was *primâ facie* evidence, not of their similarity,
but of their difference of age; and holding as he did
the doctrine of single specific centres, the conclusion
was as legitimate as any other; for the two districts
must have been occupied by migration from one of the
two, or from an intermediate spot, and the chances
against exact coincidence of migration and of imbedding
are infinite.

In point of fact, however, whether the hypothesis of single or of multiple specific centres be adopted, similarity of organic contents cannot possibly afford any proof of the synchrony of the deposits which contain them; on the contrary, it is demonstrably compatible with the lapse of the most prodigious intervals of time, and with interposition of vast changes in the organic and inorganic worlds, between the epochs in which such deposits were formed.

On what amount of similarity of their faunæ is the doctrine of the contemporaneity of the European and of the North American Silurians based? In the last edition of Sir Charles Lyell's "Elementary Geology" it is stated, on the authority of a former President of this Society, the late Daniel Sharpe, that between 30 and 40 per cent. of the species of Silurian Mollusca are common to both sides of the Atlantic. By way of due allowance for further discovery, let us double the lesser number and suppose that 60 per cent. of the species are common to the North American and the British Silurians. Sixty per cent. of species in common is, then, proof of contemporaneity.

Now suppose that, a million or two of years hence, when Britain has made another dip beneath the sea and has come up again, some geologist applies this doctrine, in comparing the strata laid bare by the upheaval of the bottom, say, of St. George's Channel with what may then remain of the Suffolk Crag. Reasoning in the same way, he will at once decide the Suffolk Crag and the St. George's Channel beds to be contemporaneous; although we happen to know that a vast period (even in the geological sense) of time, and physical changes of almost unprecedented extent, separate the two.

But if it be a demonstrable fact that strata containing more than 60 or 70 per cent. of species of Mollusca in common, and comparatively close together, may yet be separated by an amount of geological time sufficient to allow of some of the greatest physical changes the world has seen, what becomes of that sort of contemporaneity the sole evidence of which is a similarity of facies, or the identity of half a dozen species, or of a good many genera?

And yet there is no better evidence for the contemporaneity assumed by all who adopt the hypotheses of universal faunæ and floræ, of a universally uniform climate, and of a sensible cooling of the globe during geological time.

There seems, then, no escape from the admission that neither physical geology, nor palæontology, possesses any method by which the absolute synchronism of two strata can be demonstrated. All that geology can prove is local order of succession. It is mathematically certain that, in any given vertical linear section of an undisturbed series of sedimentary deposits, the bed which lies lowest is the oldest. In any other vertical linear section of the same series, of course, corresponding beds will occur in a similar order; but, however great may be the probability, no man can say with absolute certainty that the beds in the two sections were synchronously deposited. For areas of moderate extent, it is doubtless true that no practical evil is likely to result from assuming the corresponding beds to be synchronous or strictly contemporaneous; and there

are multitudes of accessory circumstances which may
fully justify the assumption of such synchrony. But
the moment the geologist has to deal with large areas,
or with completely separated deposits, the mischief
of confounding that "homotaxis" or "similarity of
arrangement," which *can* be demonstrated, with "synchrony" or "identity of date," for which there is not
a shadow of proof, under the one common term of
"contemporaneity" becomes incalculable, and proves
the constant source of gratuitous speculations.

For anything that geology or palæontology are able
to show to the contrary, a Devonian fauna and flora
in the British Islands may have been contemporaneous
with Silurian life in North America, and with a Carboniferous fauna and flora in Africa. Geographical
provinces and zones may have been as distinctly marked
in the Palæozoic epoch as at present, and those
seemingly sudden appearances of new genera and species,
which we ascribe to new creation, may be simple results
of migration.

It may be so ; it may be otherwise. In the present
condition of our knowledge and of our methods, one
verdict—"not proven, and not proveable"—must be
recorded against all the grand hypotheses of the palæontologist respecting the general succession of life on the
globe. The order and nature of terrestrial life, as a
whole, are open questions. Geology at present provides
us with most valuable topographical records, but she
has not the means of working them up into a universal
history. Is such a universal history, then, to be regarded
as unattainable ? Are all the grandest and most interesting problems which offer themselves to the

geological student essentially insoluble? Is he in the position of a scientific Tantalus—doomed always to thirst for a knowledge which he cannot obtain? The reverse is to be hoped; nay, it may not be impossible to indicate the source whence help will come.

In commencing these remarks, mention was made of the great obligations under which the naturalist lies to the geologist and palæontologist. Assuredly the time will come when these obligations will be repaid tenfold, and when the maze of the world's past history, through which the pure geologist and the pure palæontologist find no guidance, will be securely threaded by the clue furnished by the naturalist.

All who are competent to express an opinion on the subject are, at present, agreed that the manifold varieties of animal and vegetable form have not either come into existence by chance, nor result from capricious exertions of creative power; but that they have taken place in a definite order, the statement of which order is what men of science term a natural law. Whether such a law is to be regarded as an expression of the mode of operation of natural forces, or whether it is simply a statement of the manner in which a supernatural power has thought fit to act, is a secondary question, so long as the existence of the law and the possibility of its discovery by the human intellect are granted. But he must be a half-hearted philosopher who, believing in that possibility, and having watched the gigantic strides of the biological sciences during the last twenty years, doubts that science will sooner or later make this further step, so as to become possessed of the law of evolution of organic forms—of the unvarying order of that great chain of causes and effects of which all organic forms, ancient and modern, are the links. And then, if ever, we shall be able to begin to discuss, with profit, the questions respecting the commencement of life, and the nature of the successive populations of the globe, which so many seem to think are already answered.

The preceding arguments make no particular claim to novelty; indeed they have been floating more or less distinctly before the minds of geologists for the last thirty years; and if, at the present time, it has seemed desirable to give them more definite and systematic expression, it is because palæontology is every day assuming a greater importance, and now requires to rest on a basis the firmness of which is thoroughly well assured. Among its fundamental conceptions, there must be no confusion between what is certain and what is more or less probable.[1] But, pending the construction of a surer foundation than palæontology now possesses, it may be instructive, assuming for the nonce the general correctness of the ordinary hypothesis of geological contemporaneity, to consider whether the deductions which are ordinarily drawn from the whole body of palæontological facts are justifiable.

The evidence on which such conclusions are based is of two kinds, negative and positive. The value of negative evidence, in connexion with this inquiry, has been so fully and clearly discussed in an address from the chair of this Society,[2] which none of us have

---

[1] "Le plus grand service qu'on puisse rendre à la science est d'y faire place nette avant d'y rien construire."—CUVIER.

[2] Anniversary Address for 1851, Quart. Journ. Geol. Soc. vol. vii.

forgotten, that nothing need at present be said about it; the more, as the considerations which have been laid before you have certainly not tended to increase your estimation of such evidence. It will be preferable to turn to the positive facts of palæontology, and to inquire what they tell us.

We are all accustomed to speak of the number and the extent of the changes in the living population of the globe during geological time as something enormous; and indeed they are so, if we regard only the negative differences which separate the older rocks from the more modern, and if we look upon specific and generic changes as great changes, which from one point of view they truly are. But leaving the negative differences out of consideration, and looking only at the positive data furnished by the fossil world from a broader point of view—from that of the comparative anatomist who has made the study of the greater modifications of animal form his chief business—a surprise of another kind dawns upon the mind; and under *this* aspect the smallness of the total change becomes as astonishing as was its greatness under the other.

There are two hundred known orders of plants; of these not one is certainly known to exist exclusively in the fossil state. The whole lapse of geological time has as yet yielded not a single new ordinal type of vegetable structure.[1]

The positive change in passing from the recent to the ancient animal world is greater, but still singularly small. No fossil animal is so distinct from those now living as to require to be arranged even in a separate

[1] See Hooker's "Introductory Essay to the Flora of Tasmania," p. xxiii.

class from those which contain existing forms. It is only when we come to the orders, which may be roughly estimated at about a hundred and thirty, that we meet with fossil animals so distinct from those now living as to require orders for themselves; and these do not amount, on the most liberal estimate, to more than about 10 per cent. of the whole.

There is no certainly known extinct order of Protozoa; there is but one among the Cœlenterata—that of the rugose corals; there is none among the Mollusca; there are three, the Cystidea, Blastoidea, and Edrioasterida, among the Echinoderms; and two, the Trilobita and Eurypterida, among the Crustacea; making altogether five for the great sub-kingdom of Annulosa. Among Vertebrates there is no ordinally distinct fossil fish : there is only one extinct order of Amphibia—the Labyrinthodonts; but there are at least four distinct orders of Reptilia, viz. the Ichthyosauria, Plesiosauria, Pterosauria, Dinosauria, and perhaps another or two. There is no known extinct order of Birds, and no certainly known extinct order of Mammals, the ordinal distinctness of the "Toxodontia" being doubtful.

The objection that broad statements of this kind, after all, rest largely on negative evidence is obvious, but it has less force than may at first be supposed; for, as might be expected from the circumstances of the case, we possess more abundant positive evidence regarding Fishes and marine Mollusks than respecting any other forms of animal life; and yet these offer us, through the whole range of geological time, no species ordinarily distinct from those now living; while the far less numerous class of Echinoderms presents three, and the

Crustacea two, such orders, though none of these come down later than the Palæozoic age. Lastly, the Reptilia present the extraordinary and exceptional phænomenon of as many extinct as existing orders, if not more; the four mentioned maintaining their existence from the Lias to the Chalk inclusive.

Some years ago one of your Secretaries pointed out another kind of positive palæontological evidence tending towards the same conclusion—afforded by the existence of what he termed "persistent types" of vegetable and of animal life.[1] He stated, on the authority of Dr. Hooker, that there are Carboniferous plants which appear to be generically identical with some now living; that the cone of the Oolitic *Araucaria* is hardly distinguishable from that of an existing species; that a true *Pinus* appears in the Purbecks and a *Juglans* in the Chalk; while, from the Bagshot Sands, a *Banksia*, the wood of which is not distinguishable from that of species now living in Australia, had been obtained.

Turning to the animal kingdom, he affirmed the tabulate corals of the Silurian rocks to be wonderfully like those which now exist; while even the families of the Aporosa were all represented in the older Mesozoic rocks.

Among the Mollusca similar facts were adduced. Let it be borne in mind that *Avicula, Mytilus, Chiton, Natica, Patella, Trochus, Discina, Orbicula, Lingula, Rhynchonella,* and *Nautilus,* all of which are existing *genera,* are given without a doubt as Silurian in the

---

[1] See the abstract of a Lecture "On the Persistent Types of Animal Life" in the "Notices of the Meetings of the Royal Institution of Great Britain," June 3, 1859, vol. iii. p. 151.

last edition of " Siluria ; " while the highest forms of the highest Cephalopods are represented in the Lias by a genus, *Belemnoteuthis*, which presents the closest relation to the existing *Loligo*.

The two highest groups of the Annulosa, the Insecta and the Arachnida, are represented in the Coal, either by existing genera, or by forms differing from existing genera in quite minor peculiarities.

Turning to the Vertebrata, the only palæozoic Elasmobranch Fish of which we have any complete knowledge is the Devonian and Carboniferous *Pleuracanthus*, which differs no more from existing Sharks than these do from one another.

Again, vast as is the number of undoubtedly Ganoid fossil Fishes, and great as is their range in time, a large mass of evidence has recently been adduced to show that almost all those respecting which we possess sufficient information, are referable to the same sub-ordinal groups as the existing *Lepidosteus, Polypterus*, and Sturgeon ; and that a singular relation obtains between the older and the younger Fishes ; the former, the Devonian Ganoids, being almost all members of the same sub-order as *Polypterus*, while the Mesozoic Ganoids are almost all similarly allied to *Lepidosteus*.[1]

Again, what can be more remarkable than the singular constancy of structure preserved throughout a vast period of time by the family of the Pycnodonts and by that of the true Cœlacanths : the former persisting, with but insignificant modifications, from the Carboniferous to the

[1] " Memoirs of the Geological Survey of the United Kingdom.—Decade x. Preliminary Essay upon the Systematic Arrangement of the Fishes of the Devonian Epoch."

Tertiary rocks, inclusive ; the latter existing, with still less change, from the Carboniferous rocks to the Chalk, inclusive ?

Among Reptiles, the highest living group, that of the Crocodilia, is represented, at the early part of the Mesozoic epoch, by species identical in the essential characters of their organization with those now living, and differing from the latter only in such matters as the form of the articular facets of the vertebral centra, in the extent to which the nasal passages are separated from the cavity of the mouth by bone, and in the proportions of the limbs.

And even as regards the Mammalia, the scanty remains of Triassic and Oolitic species afford no foundation for the supposition that the organization of the oldest forms differed nearly so much from some of those which now live as these differ from one another.

It is needless to multiply these instances ; enough has been said to justify the statement that, in view of the immense diversity of known animal and vegetable forms, and the enormous lapse of time indicated by the accumulation of fossiliferous strata, the only circumstance to be wondered at is, not that the changes of life, as exhibited by positive evidence, have been so great, but that they have been so small.

Be they great or small, however, it is desirable to attempt to estimate them. Let us, therefore, take each great division of the animal world in succession, and, whenever an order or a family can be shown to have had a prolonged existence, let us endeavour to ascertain how far the later members of the group differ from the

earlier ones.   If these later members, in all or in many
cases, exhibit a certain amount of modification, the fact
is, so far, evidence in favour of a general law of change ;
and, in a rough way, the rapidity of that change will be
measured by the demonstrable amount of modification.
On the other hand, it must be recollected that the
absence of any modification, while it may leave the
doctrine of the existence of a law of change without
positive support, cannot possibly disprove all forms of
that doctrine, though it may afford a sufficient refutation of many of them.

The PROTOZOA.—The Protozoa are represented throughout the whole range of geological series, from the Lower
Silurian formation to the present day.   The most
ancient forms recently made known by Ehrenberg are
exceedingly like those which now exist : no one has
ever pretended that the difference between any ancient
and any modern Foraminifera is of more than generic
value ; nor are the oldest Foraminifera either simpler,
more embryonic, or less differentiated, than the existing
forms.

The CŒLENTERATA.—The Tabulate Corals have existed
from the Silurian epoch to the present day, but I am not
aware that the ancient *Heliolites* possesses a single mark
of a more embryonic or less differentiated character, or
less high organization, than the existing *Heliopora*.   As
for the Aporose Corals, in what respect is the Silurian
Palæocyclus less highly organized or more embryonic
than the modern *Fungia*, or the Liassic Aporosa than
the existing members of the same families ?

The *Mollusca.*—In what sense is the living *Wald-
heimia* less embryonic, or more specialized, than the

palæozoic *Spirifer ;* or the existing *Rhynchonellæ, Craniæ, Discinæ, Lingulæ,* than the Silurian species of the same genera? In what sense can *Loligo* or *Spirula* be said to be more specialized, or less embryonic, than *Belemnites;* or the modern species of Lamellibranch and Gasteropod genera, than the Silurian species of the same genera?

The ANNULOSA.—The Carboniferous Insecta and Arachnida are neither less specialized, nor more embryonic, than those that now live, nor are the Liassic Cirripedia and Macrura; while several of the Brachyura, which appear in the Chalk, belong to existing genera; and none exhibit either an intermediate, or an embryonic, character.

The VERTEBRATA.—Among fishes I have referred to the Cœlacanthini (comprising the genera *Cœlacanthus, Holophagus, Undina,* and *Macropoma*) as affording an example of a persistent type; and it is most remarkable to note the smallness of the differences between any of these fishes (affecting at most the proportions of the body and fins, and the character and sculpture of the scales), notwithstanding their enormous range in time. In all the essentials of its very peculiar structure, the *Macropoma* of the Chalk is identical with the *Cœlacanthus* of the Coal. Look at the genus *Lepidotus,* again, persisting without a modification of importance from the Liassic to the Eocene formations, inclusive.

Or among the Teleostei—in what respect is the *Beryx* of the Chalk more embryonic, or less differentiated, than *Beryx lineatus* of King George's Sound?

Or to turn to the higher Vertebrata—in what sense are the Liassic Chelonia inferior to those which now exist? How are the Cretaceous Ichthyosauria, Plesiosauria, or Pterosauria less embryonic, or more differentiated, species than those of the Lias?

Or lastly, in what circumstance is the *Phascolotherium* more embryonic, or of a more generalized type, than the modern Opossum; or a *Lophiodon*, or a *Palæotherium*, than a modern *Tapirus* or *Hyrax?*

These examples might be almost indefinitely multiplied, but surely they are sufficient to prove that the only safe and unquestionable testimony we can procure —positive evidence—fails to demonstrate any sort of progressive modification towards a less embryonic, or less generalized, type in a great many groups of animals of long-continued geological existence. In these groups there is abundant evidence of variation—none of what is ordinarily understood as progression; and, if the known geological record is to be regarded as even any considerable fragment of the whole, it is inconceivable that any theory of a necessarily progressive development can stand, for the numerous orders and families cited afford no trace of such a process.

But it is a most remarkable fact, that, while the groups which have been mentioned, and many besides, exhibit no sign of progressive modification, there are others, coexisting with them, under the same conditions, in which more or less distinct indications of such a process seem to be traceable. Among such indications I may remind you of the predominance of Holostome Gasteropoda in the older rocks as compared with that of Siphonostome Gasteropoda in the later. A case less open to the objection of negative evidence, however, is that afforded by the Tetrabranchiate Cephalopoda, the forms of the shells and of the septal sutures exhibiting a certain increase of complexity in the newer genera. Here, however, one is met at once with the occurrence of *Orthoceras* and *Baculites* at the two ends of the series, and of the fact that one of the simplest genera, *Nautilus*, is that which now exists.

The Crinoidea, in the abundance of stalked forms in the ancient formations as compared with their present rarity, seem to present us with a fair case of modification from a more embryonic towards a less embryonic condition. But then, on careful consideration of the facts, the objection arises that the stalk, calyx, and arms of the palæozoic Crinoid are exceedingly different from the corresponding organs of a larval *Comatula;* and it might with perfect justice be argued that *Actinocrinus* and *Eucalyptocrinus*, for example, depart to the full as widely, in one direction, from the stalked embryo of *Comatula*, as *Comatula* itself does in the other.

The Echinidea, again, are frequently quoted as exhibiting a gradual passage from a more generalized to a more specialized type, seeing that the elongated, or oval, Spatangoids appear after the spheroidal Echinoids. But here it might be argued, on the other hand, that the spheroidal Echinoids, in reality, depart further from the general plan and from the embryonic form than the elongated Spatangoids do; and that the peculiar dental apparatus and the pedicellariæ of the former are marks of at least as great differentiation as the petaloid ambulacra and semitæ of the latter.

Once more, the prevalence of Macrurous before Brachyurous Podophthalmia is, apparently, a fair piece of evidence in favour of progressive modification in the same order of Crustacea; and yet the case will not stand much sifting, seeing that the Macrurous Podophthalmia depart as far in one direction from the common type of Podophthalmia, or from any embryonic condition of the Brachyura, as the Brachyura do in the other; and that the middle terms between Macrura and Brachyura — the Anomura — are little better represented in the older Mesozoic rocks than the Brachyura are.

None of the cases of progressive modification which are cited from among the Invertebrata appear to me to have a foundation less open to criticism than these; and if this be so, no careful reasoner would, I think, be inclined to lay very great stress upon them. Among the Vertebrata, however, there are a few examples which appear to be far less open to objection.

It is, in fact, true of several groups of Vertebrata which have lived through a considerable range of time, that the endoskeleton (more particularly the spinal column) of the older genera presents a less ossified, and, so far, less differentiated, condition than that of the younger genera. Thus the Devonian Ganoids, though almost all members of the same sub-order as *Polypterus*, and presenting numerous important resemblances to the existing genus, which possesses biconcave vertebræ, are, for the most part, wholly devoid of ossified vertebral centra. The Mesozoic Lepidosteidæ, again, have, at most, biconcave vertebræ, while the existing *Lepidosteus* has Salamandroid, opisthocœlous, vertebræ. So, none of the Palæozoic Sharks have shown themselves to be possessed of ossified vertebræ, while the majority of modern Sharks possess such vertebræ. Again, the more ancient

Crocodilia and Lacertilia have vertebræ with the articular facets of their centra flattened or biconcave, while the modern members of the same group have them proceolous. But the most remarkable examples of progressive modification of the vertebral column, in correspondence with geological age, are those afforded by the Pycnodonts among fish, and the Labyrinthodonts among Amphibia.

The late able ichthyologist Heckel pointed out the fact, that, while the Pycnodonts never possess true vertebral centra, they differ in the degree of expansion and extension of the ends of the bony arches of the vertebræ upon the sheath of the notochord; the Carboniferous forms exhibiting hardly any such expansion, while the Mesozoic genera present a greater and greater development, until, in the Tertiary forms, the expanded ends become suturally united so as to form a sort of false vertebra. Hermann von Meyer, again, to whose luminous researches we are indebted for our present large knowledge of the organization of the older Labyrinthodonts, has proved that the Carboniferous *Archegosaurus* had very imperfectly developed vertebral centra, while the Triassic *Mastodonsaurus* had the same parts completely ossified.[1]

The regularity and evenness of the dentition of the *Anoplotherium*, as contrasted with that of existing Artiodactyles, and the assumed nearer approach of the dentition of certain ancient Carnivores to the typical arrangement, have also been cited as exemplifications of

[1] As this Address is passing through the press (March 7, 1862), evidence lies before me of the existence of a new Labyrinthodont (*Pholidogaster*), from the Edinburgh coal-field, with well-ossified vertebral centra.

a law of progressive development, but I know of no other cases based on positive evidence which are worthy of particular notice.

What then does an impartial survey of the positively ascertained truths of palæontology testify in relation to the common doctrines of progressive modification, which suppose that modification to have taken place by a necessary progress from more to less embryonic forms, or from more to less generalized types, within the limits of the period represented by the fossiliferous rocks?

It negatives those doctrines; for it either shows us no evidence of any such modification, or demonstrates it to have been very slight; and as to the nature of that modification, it yields no evidence whatsoever that the earlier members of any long-continued group were more generalized in structure than the later ones. To a certain extent, indeed, it may be said that imperfect ossification of the vertebral column is an embryonic character; but, on the other hand, it would be extremely incorrect to suppose that the vertebral columns of the older Vertebrata are in any sense embryonic in their whole structure.

Obviously, if the earliest fossiliferous rocks now known are coeval with the commencement of life, and if their contents give us any just conception of the nature and the extent of the earliest fauna and flora, the insignificant amount of modification which can be demonstrated to have taken place in any one group of animals, or plants, is quite incompatible with the hypothesis that all living forms are the results of a necessary process of progressive development, entirely comprised within the time represented by the fossiliferous rocks.

Contrariwise, any admissible hypothesis of progressive modification must be compatible with persistence without progression, through indefinite periods. And should such an hypothesis eventually be proved to be true, in the only way in which it can be demonstrated, viz. by observation and experiment upon the existing forms of life, the conclusion will inevitably present itself, that the Palæozoic, Mesozoic, and Cainozoic faunæ and floræ, taken together, bear somewhat the same proportion to the whole series of living beings which have occupied this globe, as the existing fauna and flora do to them.

Such are the results of palæontology as they appear, and have for some years appeared, to the mind of an inquirer who regards that study simply as one of the applications of the great biological sciences, and who desires to see it placed upon the same sound basis as other branches of physical inquiry. If the arguments which have been brought forward are valid, probably no one, in view of the present state of opinion, will be inclined to think the time wasted which has been spent upon their elaboration.

XI.

GEOLOGICAL REFORM.

"A great reform in geological speculation seems now to have become necessary."

"It is quite certain that a great mistake has been made,—that British popular geology at the present time is in direct opposition to the principles of Natural Philosophy."[1]

In reviewing the course of geological thought during the past year, for the purpose of discovering those matters to which I might most fitly direct your attention in the Address which it now becomes my duty to deliver from the Presidential Chair, the two somewhat alarming sentences which I have just read, and which occur in an able and interesting essay by an eminent natural philosopher, rose into such prominence before my mind that they eclipsed everything else.

It surely is a matter of paramount importance for the British geologists (some of them very popular geologists too) here in solemn annual session assembled, to inquire whether the severe judgment thus passed upon them by so high an authority as Sir William Thomson is one to

---

[1] On Geological Time. By Sir W. Thomson, LL.D. Transactions of the Geological Society of Glasgow, vol. iii.

which they must plead guilty *sans phrase,* or whether they are prepared to say " not guilty," and appeal for a reversal of the sentence to that higher court of educated scientific opinion to which we are all amenable.

As your attorney-general for the time being, I thought I could not do better than get up the case with a view of advising you.   It is true that the charges brought forward by the other side involve the consideration of matters quite foreign to the pursuits with which I am ordinarily occupied ; but, in that respect, I am only in the position which is, nine times out of ten, occupied by counsel, who nevertheless contrive to gain their causes, mainly by force of mother-wit and common sense, aided by some training in other intellectual exercises.

Nerved by such precedents, I proceed to put my pleading before you.

And the first question with which I propose to deal is, What is it to which Sir W. Thomson refers when he speaks of "geological speculation" and "British popular geology"?

I find three, more or less contradictory, systems of geological thought, each of which might fairly enough claim these appellations, standing side by side in Britain. I shall call one of them CATASTROPHISM, another UNIFORMITARIANISM, the third EVOLUTIONISM ; and I shall try briefly to sketch the characters of each, that you may say whether the classification is, or is not, exhaustive.

By CATASTROPHISM, I mean any form of geological speculation which, in order to account for the phænomena of geology, supposes the operation of forces different in their nature, or immeasurably different in power, from those which we at present see in action in the universe.

The Mosaic cosmogony is, in this sense, catastrophic, because it assumes the operation of extra-natural power. The doctrine of violent upheavals, *débâcles*, and cataclysms in general, is catastrophic, so far as it assumes that these were brought about by causes which have now no parallel. There was a time when catastrophism might, pre-eminently, have claimed the title of " British popular geology;" and assuredly it has yet many adherents, and reckons among its supporters some of the most honoured members of this Society.

By UNIFORMITARIANISM, I mean especially, the teaching of Hutton and of Lyell.

That great, though incomplete work, " The Theory of the Earth," seems to me to be one of the most remarkable contributions to geology which is recorded in the annals of the science. So far as the not-living world is concerned, uniformitarianism lies there, not only in germ, but in blossom and fruit.

If one asks how it is that Hutton was led to entertain views so far in advance of those prevalent in his time, in some respects ; while, in others, they seem almost curiously limited, the answer appears to me to be plain.

Hutton was in advance of the geological speculation of his time, because, in the first place, he had amassed a vast store of knowledge of the facts of geology, gathered by personal observation in travels of considerable extent; and because, in the second place, he was thoroughly trained in the physical and chemical science of his day, and thus possessed, as much as any one in his time could possess it, the knowledge which is requisite for the just interpretation of geological phænomena, and the habit of thought which fits a man for scientific inquiry.

It is to this thorough scientific training, that I ascribe Hutton's steady and persistent refusal to look to other causes than those now in operation, for the explanation of geological phænomena.

Thus he writes :—" I do not pretend, as he [M. de Luc] does in his theory, to describe the beginning of things. I take things such as I find them at present ; and from these I reason with regard to that which must have been." [1]

And again :—" A theory of the earth, which has for object truth, can have no retrospect to that which had preceded the present order of the world ; for this order alone is what we have to reason upon ; and to reason without data is nothing but delusion. A theory, there-fore, which is limited to the actual constitution of this earth cannot be allowed to proceed one step beyond the present order of things." [2]

And so clear is he, that no causes beside such as are now in operation are needed to account for the character and disposition of the components of the crust of the earth, that he says, broadly and boldly :—" . . . There is no part of the earth which has not had the same origin, so far as this consists in that earth being collected at the bottom of the sea, and afterwards produced, as land, along with masses of melted substances, by the operation of mineral causes." [3]

But other influences were at work upon Hutton beside those of a mind logical by Nature, and scientific by sound training ; and the peculiar turn which his specu-lations took seems to me to be unintelligible, unless these

---

[1] The Theory of the Earth, vol. i. p. 173, note.   [2] Ibid. p. 281.
[3] Ibid. p. 371.

be taken into account. The arguments of the French astronomers and mathematicians, which, at the end of the last century, were held to demonstrate the existence of a compensating arrangement among the celestial bodies, whereby all perturbations eventually reduced themselves to oscillations on each side of a mean position, and the stability of the solar system was secured, had evidently taken strong hold of Hutton's mind.

In those oddly constructed periods which seem to have prejudiced many persons against reading his works, but which are full of that peculiar, if unattractive, eloquence which flows from mastery of the subject, Hutton says :—

"We have now got to the end of our reasoning ; we have no data further to conclude immediately from that which actually is. But we have got enough ; we have the satisfaction to find, that in Nature there is wisdom, system, and consistency. For having, in the natural history of this earth, seen a succession of worlds, we may from this conclude that there is a system in Nature ; in like manner as, from seeing revolutions of the planets, it is concluded, that there is a system by which they are intended to continue those revolutions. But if the succession of worlds is established in the system of Nature, it is in vain to look for anything higher in the origin of the earth. The result, therefore, of this physical inquiry is, that we find no vestige of a beginning,—no prospect of an end." [1]

Yet another influence worked strongly upon Hutton. Like most philosophers of his age, he coquetted with those final causes which have been named barren virgins, but which might be more fitly termed the *hetairæ* of

[1] The Theory of the Earth, vol. i. p. 200.

philosophy, so constantly have they led men astray. The final cause of the existence of the world is, for Hutton, the production of life and intelligence.

"We have now considered the globe of this earth as a machine, constructed upon chemical as well as mechanical principles, by which its different parts are all adapted, in form, in quality, and in quantity, to a certain end ; an end attained with certainty or success ; and an end from which we may perceive wisdom, in contemplating the means employed.

"But is this world to be considered thus merely as a machine, to last no longer than its parts retain their present position, their proper forms and qualities ? Or may it not be also considered as an organized body ? such as has a constitution in which the necessary decay of the machine is naturally repaired, in the exertion of those productive powers by which it had been formed.

"This is the view in which we are now to examine the globe ; to see if there be, in the constitution of this world, a reproductive operation, by which a ruined constitution may be again repaired, and a duration or stability thus procured to the machine, considered as a world sustaining plants and animals." [1]

Kirwan, and the other Philistines of the day, accused Hutton of declaring that his theory implied that the world never had a beginning, and never differed in condition from its present state. Nothing could be more grossly unjust, as he expressly guards himself against any such conclusion in the following terms :—

"But in thus tracing back the natural operations which have succeeded each other, and mark to us the

[1] The Theory of the Earth, vol. i. pp. 16, 17.

course of time past, we come to a period in which we cannot see any farther. This, however, is not the beginning of the operations which proceed in time and according to the wise economy of this world ; nor is it the establishing of that which, in the course of time, had no beginning ; it is only the limit of our retrospective view of those operations which have come to pass in time, and have been conducted by supreme intelligence." [1]

I have spoken of Uniformitarianism as the doctrine of Hutton and of Lyell. If I have quoted the older writer rather than the newer, it is because his works are little known, and his claims on our veneration too frequently forgotten, not because I desire to dim the fame of his eminent successor. Few of the present generation of geologists have read Playfair's "Illustrations," fewer still the original "Theory of the Earth ;" the more is the pity ; but which of us has not thumbed every page of the "Principles of Geology?" I think that he who writes fairly the history of his own progress in geological thought, will not be able to separate his debt to Hutton from his obligations to Lyell; and the history of the progress of individual geologists is the history of geology.

No one can doubt that the influence of uniformitarian views has been enormous, and, in the main, most beneficial and favourable to the progress of sound geology.

Nor can it be questioned that Uniformitarianism has even a stronger title than Catastrophism to call itself the geological speculation of Britain, or, if you will, British popular geology. For it is eminently a British doctrine, and has even now made comparatively little progress

---

[1] The Theory of the Earth, vol. i. p. 223.

on the continent of Europe. Nevertheless it seems to me to be open to serious criticism upon one of its aspects.

I have shown how unjust was the insinuation that Hutton denied a beginning to the world. But it would not be unjust to say that he persistently, in practice, shut his eyes to the existence of that prior and different state of things which, in theory, he admitted; and, in this aversion to look beyond the veil of stratified rocks, Lyell follows him.

Hutton and Lyell alike agree in their indisposition to carry their speculations a step beyond the period recorded in the most ancient strata now open to observation in the crust of the earth. This is, for Hutton, "the point in which we cannot see any farther;" while Lyell tells us,—

"The astronomer may find good reasons for ascribing the earth's form to the original fluidity of the mass, in times long antecedent to the first introduction of living beings into the planet; but the geologist must be content to regard the earliest monuments which it is his task to interpret, as belonging to a period when the crust had already acquired great solidity and thickness, probably as great as it now possesses, and when volcanic rocks, not essentially differing from those now produced, were formed from time to time, the intensity of volcanic heat being neither greater nor less than it is now." [1]

And again, "As geologists, we learn that it is not only the present condition of the globe which has been suited to the accommodation of myriads of living creatures, but that many former states also have been adapted to the

[1] Principles of Geology, vol. ii. p. 211.

organization and habits of prior races of beings. The
disposition of the seas, continents and islands, and the
climates, have varied; the species likewise have been
changed; and yet they have all been so modelled, on
types analogous to those of existing plants and animals,
as to indicate, throughout, a perfect harmony of design,
and unity of purpose. To assume that the evidence of
the beginning, or end, of so vast a scheme lies within
the reach of our philosophical inquiries, or even of our
speculations, appears to be inconsistent with a just
estimate of the relations which subsist between the finite
powers of man and the attributes of an infinite and
eternal Being." [1]

The limitations implied in these passages appear to
me to constitute the weakness and the logical defect of
uniformitarianism. No one will impute blame to Hutton
that, in face of the imperfect condition, in his day, of
those physical sciences which furnish the keys to the
riddles of geology, he should have thought it practical
wisdom to limit his theory to an attempt to account for
" the present order of things; " but I am at a loss to comprehend why, for all time, the geologist must be content
to regard the oldest fossiliferous rocks as the *ultima
Thule* of his science; or what there is inconsistent with
the relations between the finite and the infinite mind, in
the assumption, that we may discern somewhat of the
beginning, or of the end, of this speck in space we call
our earth. The finite mind is certainly competent to
trace out the development of the fowl within the egg;
and I know not on what ground it should find more
difficulty in unravelling the complexities of the develop-

---

[1] Principles of Geology, vol. ii. p. 613.

ment of the earth.   In fact, as Kant has well remarked,[1] the cosmical process is really simpler than the biological.

This attempt to limit, at a particular point, the progress of inductive and deductive reasoning from the things which are, to those which were—this faithlessness to its own logic, seems to me to have cost Uniformitarianism the place, as the permanent form of geological specula- tion, which it might otherwise have held.

It remains that I should put before you what I understand to be the third phase of geological specula- tion—namely, EVOLUTIONISM.

I shall not make what I have to say on this head clear, unless I diverge, or seem to diverge, for a while, from the direct path of my discourse, so far as to explain what I take to be the scope of geology itself.   I conceive geology to be the history of the earth, in precisely the same sense as biology is the history of living beings; and I trust you will not think that I am overpowered by the influence of a dominant pursuit if I say that I trace a close analogy between these two histories.

If I study a living being, under what heads does the knowledge I obtain fall?   I can learn its structure, or what we call its ANATOMY ; and its DEVELOPMENT, or the series of changes which it passes through to acquire its complete structure.   Then I find that the living being has certain powers resulting from its own acti- vities, and the interaction of these with the activities of

---

[1] " Man darf es sich also nicht befremden lassen, wenn ich mich unterstehe zu sagen, dass eher die Bildung aller Himmelskörper, die Ursache ihrer Bewegungen, kurz der Ursprung der ganzen gegenwärtigen Verfassung des Weltbaues werden können eingesehen werden, ehe die Erzeugung eines einzigen Krautes oder einer Raupe aus mechanischen Gründen, deutlich und vollständig kund werden wird."—KANT's *Sämmtliche Werke*, Bd. I. p. 220.

other things—the knowledge of which is PHYSIOLOGY.
Beyond this the living being has a position in space and
time, which is its DISTRIBUTION. All these form the
body of ascertainable facts which constitute the *status
quo* of the living creature. But these facts have their
causes; and the ascertainment of these causes is the
doctrine of ÆTIOLOGY.

If we consider what is knowable about the earth, we
shall find that such earth-knowledge—if I may so trans-
late the word geology—falls into the same categories.

What is termed stratigraphical geology is neither more
nor less than the anatomy of the earth; and the history
of the succession of the formations is the history of a
succession of such anatomies, or corresponds with development, as distinct from generation.

The internal heat of the earth, the elevation and
depression of its crust, its belchings forth of vapours,
ashes, and lava, are its activities, in as strict a sense, as are
warmth and the movements and products of respiration
the activities of an animal. The phænomena of the
seasons, of the trade winds, of the Gulf-stream, are as
much the results of the reaction between these inner
activities and outward forces, as are the budding of the
leaves in spring and their falling in autumn the effects
of the interaction between the organization of a plant
and the solar light and heat. And, as the study of the
activities of the living being is called its physiology, so
are these phænomena the subject-matter of an analogous
telluric physiology, to which we sometimes give the
name of meteorology, sometimes that of physical geography, sometimes that of geology. Again, the earth
has a place in space and in time, and relations to other

bodies in both these respects, which constitute its distribution. This subject is usually left to the astronomer;
but a knowledge of its broad outlines seems to me to be
an essential constituent of the stock of geological ideas.

All that can be ascertained concerning the structure,
succession of conditions, actions, and position in space of
the earth, is the matter of fact of its natural history.
But, as in biology, there remains the matter of reasoning
from these facts to their causes, which is just as much
science as the other, and indeed more; and this constitutes geological ætiology.

Having regard to this general scheme of geological
knowledge and thought, it is obvious that geological
speculation may be, so to speak, anatomical and developmental speculation, so far as it relates to points of stratigraphical arrangement which are out of reach of direct
observation; or, it may be physiological speculation, so
far as it relates to undetermined problems relative to the
activities of the earth; or, it may be distributional speculation, if it deals with modifications of the earth's place
in space; or, finally, it will be ætiological speculation, if
it attempts to deduce the history of the world, as a
whole, from the known properties of the matter of the
earth, in the conditions in which the earth has been placed.

For the purposes of the present discourse I may take
this last to be what is meant by "geological speculation."

Now uniformitarianism, as we have seen, tends to
ignore geological speculation in this sense altogether.

The one point the catastrophists and the uniformi-
tarians agreed upon, when this Society was founded, was
to ignore it. And you will find, if you look back into
our records, that our revered fathers in geology plumed

themselves a good deal upon the practical sense and wisdom of this proceeding. As a temporary measure, I do not presume to challenge its wisdom ; but in all organized bodies temporary changes are apt to produce permanent effects ; and as time has slipped by, altering all the conditions which may have made such mortification of the scientific flesh desirable, I think the effect of the stream of cold water which has steadily flowed over geological speculation within these walls, has been of doubtful beneficence.

The sort of geological speculation to which I am now referring (geological ætiology, in short) was created, as a science, by that famous philosopher Immanuel Kant, when, in 1755, he wrote his " General Natural History and Theory of the Celestial Bodies ; or an Attempt to account for the Constitution and the mechanical Origin of the Universe upon Newtonian principles." [1]

In this very remarkable, but seemingly little-known treatise,[2] Kant expounds a complete cosmogony, in the shape of a theory of the causes which have led to the development of the universe from diffused atoms of matter endowed with simple attractive and repulsive forces.

" Give me matter," says Kant, " and I will build the world ; " and he proceeds to deduce from the simple data from which he starts, a doctrine in all essential respects similar to the well-known " Nebular Hypothesis " of Laplace.[3] He accounts for the relation of the masses

[1] Grant ("History of Physical Astronomy," p. 574) makes but the briefest reference to Kant.

[2] " Allgemeine Naturgeschichte und Theorie des Himmels ; oder Versuch von der Verfassung und dem mechanischen Ursprunge des ganzen Weltgebäudes nach Newton'schen Grundsatzen abgehandelt."—KANT's *Sämmtliche Werke*, Bd. i. p. 207.

[3] Système du Monde, tome ii. chap. 6

and the densities of the planets to their distances from
the sun, for the eccentricities of their orbits, for their
rotations, for their satellites, for the general agreement
in the direction of rotation among the celestial bodies,
for Saturn's ring, and for the zodiacal light. He finds,
in each system of worlds, indications that the attractive
force of the central mass will eventually destroy its organization, by concentrating upon itself the matter of the
whole system ; but, as the result of this concentration,
he argues for the development of an amount of heat
which will dissipate the mass once more into a molecular
chaos such as that in which it began.

Kant pictures to himself the universe as once an
infinite expansion of formless and diffused matter. At
one point of this he supposes a single centre of attraction
set up; and, by strict deductions from admitted dynamical
principles, shows how this must result in the development
of a prodigious central body, surrounded by systems of
solar and planetary worlds in all stages of development.
In vivid language he depicts the great world-maelstrom,
widening the margins of its prodigious eddy in the slow
progress of millions of ages, gradually reclaiming more
and more of the molecular waste, and converting chaos
into cosmos. But what is gained at the margin is lost
in the centre ; the attractions of the central systems
bring their constituents together, which then, by the heat
evolved, are converted once more into molecular chaos.
Thus the worlds that are, lie between the ruins of the
worlds that have been and the chaotic materials of the
worlds that shall be ; and, in spite of all waste and
destruction, Cosmos is extending his borders at the
expense of Chaos.

Kant's further application of his views to the earth itself is to be found in his "Treatise on Physical Geography"[1] (a term under which the then unknown science of geology was included), a subject which he had studied with very great care and on which he lectured for many years. The fourth section of the first part of this Treatise is called " History of the great Changes which the Earth has formerly undergone and is still undergoing," and is, in fact, a brief and pregnant essay upon the principles of geology. Kant gives an account first " of the gradual changes which are now taking place" under the heads of such as are caused by earthquakes, such as are brought about by rain and rivers, such as are effected by the sea, such as are produced by winds and frost; and, finally, such as result from the operations of man.

The second part is devoted to the " Memorials of the Changes which the Earth has undergone in remote antiquity." These are enumerated as :—A. Proofs that the sea formerly covered the whole earth. B. Proofs that the sea has often been changed into dry land and then again into sea. C. A discussion of the various theories of the earth put forward by Scheuchzer, Moro, Bonnet, Woodward, White, Leibnitz, Linnæus, and Buffon.

The third part contains an " Attempt to give a sound explanation of the ancient history of the earth."

I suppose that it would be very easy to pick holes in the details of Kant's speculations, whether cosmological, or specially telluric, in their application. But, for all that, he seems to me to have been the first person to

[1] Kant's " Sämmtliche Werke," Bd. viii. p. 145.

frame a complete system of geological speculation by
founding the doctrine of evolution.

With as much truth as Hutton, Kant could say, " I
take things just as I find them at present, and, from
these, I reason with regard to that which must have
been." Like Hutton, he is never tired of pointing
out that " in Nature there is wisdom, system, and consistency." And, as in these great principles, so in believing that the cosmos has a reproductive operation " by
which a ruined constitution may be repaired," he fore-
stalls Hutton ; while, on the other hand, Kant is true to
science. He knows no bounds to geological speculation
but those of the intellect. He reasons back to a beginning of the present state of things ; he admits the possibility of an end.

I have said that the three schools of geological speculation which I have termed Catastrophism, Uniformitarianism, and Evolutionism are commonly supposed to
be antagonistic to one another; and I presume it will
have become obvious that, in my belief, the last is
destined to swallow up the other two. But it is proper
to remark that each of the latter has kept alive the tradition of precious truths.

CATASTROPHISM has insisted upon the existence of a
practically unlimited bank of force, on which the theorist
might draw ; and it has cherished the idea of the development of the earth from a state in which its form,
and the forces which it exerted, were very different from
those we now know. That such difference of form and
power once existed is a necessary part of the doctrine of
evolution.

UNIFORMITARIANISM, on the other hand, has with

equal justice insisted upon a practically unlimited bank
of time, ready to discount any quantity of hypothetical
paper. It has kept before our eyes the power of the
infinitely little, time being granted, and has compelled us
to exhaust known causes, before flying to the unknown.

To my mind there appears to be no sort of necessary
theoretical antagonism between Catastrophism and Uniformitarianism. On the contrary, it is very conceivable
that catastrophes may be part and parcel of uniformity.
Let me illustrate my case by analogy. The working of
a clock is a model of uniform action ; good time-keeping
means uniformity of action. But the striking of the
clock is essentially a catastrophe ; the hammer might be
made to blow up a barrel of gunpowder, or turn on a
deluge of water ; and, by proper arrangement, the clock,
instead of marking the hours, might strike at all sorts of
irregular periods, never twice alike, in the intervals,
force, or number of its blows. Nevertheless, all these
irregular, and apparently lawless, catastrophes would be
the result of an absolutely uniformitarian action ; and
we might have two schools of clock-theorists, one
studying the hammer and the other the pendulum.

Still less is there any necessary antagonism between
either of these doctrines and that of Evolution, which
embraces all that is sound in both Catastrophism and
Uniformitarianism, while it rejects the arbitrary assumptions of the one and the, as arbitrary, limitations of the
other. Nor is the value of the doctrine of Evolution to the
philosophic thinker diminished by the fact that it applies
the same method to the living and the not-living world ;
and embraces, in one stupendous analogy, the growth
of a solar system from molecular chaos, the shaping

of the earth from the nebulous cubhood of its youth, through innumerable changes and immeasurable ages, to its present form ; and the development of a living being from the shapeless mass of protoplasm we term a germ.

I do not know whether Evolutionism can claim that amount of currency which would entitle it to be called British popular geology ; but, more or less vaguely, it is assuredly present in the minds of most geologists.

Such being the three phases of geological speculation, we are now in a position to inquire which of these it is that Sir William Thomson calls upon us to reform in the passages which I have cited.

It is obviously Uniformitarianism which the distinguished physicist takes to be the representative of geological speculation in general. And thus a first issue is raised, inasmuch as many persons (and those not the least thoughtful among the younger geologists) do not accept strict Uniformitarianism as the final form of geological speculation. We should say, if Hutton and Playfair declare the course of the world to have been always the same, point out the fallacy by all means ; but, in so doing, do not imagine that you are proving modern geology to be in opposition to natural philosophy. I do not suppose that, at the present day, any geologist would be found to maintain absolute Uniformitarianism, to deny that the rapidity of the rotation of the earth *may* be diminishing, that the sun *may* be waxing dim, or that the earth itself *may* be cooling. Most of us, I suspect, are Gallios, " who care for none of these things," being of opinion that, true or fictitious, they have made no practical difference to the earth, during the period of which a record is preserved in stratified deposits.

The accusation that we have been running counter to the *principles* of natural philosophy, therefore, is devoid of foundation. The only question which can arise is whether we have, or have not, been tacitly making assumptions which are in opposition to certain conclusions which may be drawn from those principles. And this question subdivides itself into two :—the first, are we really contravening such conclusions ? the second, if we are, are those conclusions so firmly based that we may not contravene them ? I reply in the negative to both these questions, and I will give you my reasons for so doing. Sir William Thomson believes that he is able to prove, by physical reasonings, " that the existing state of things on the earth, life on the earth —all geological history showing continuity of life— must be limited within some such period of time as one hundred million years " (loc. cit. p. 25).

The first inquiry which arises plainly is, has it ever been denied that this period *may* be enough for the purposes of geology ?

The discussion of this question is greatly embarrassed by the vagueness with which the assumed limit is, I will not say defined, but indicated,—" some such period of past time as one hundred million years." Now does this mean that it may have been two, or three, or four hundred million years ? Because this really makes all the difference.[1]

---

[1] Sir William Thomson implies (loc. cit. p. 16), that the precise time is of no consequence : "the principle is the same ;" but, as the principle is admitted, the whole discussion turns on its practical results.

I presume that 100,000 feet may be taken as a full allowance for the total thickness of stratified rocks containing traces of life ; 100,000 divided by 100,000,000 = 0·001. Consequently, the deposit of 100,000 feet of stratified rock in 100,000,000 years means that the deposit has taken place at the rate of $\frac{1}{1000}$ of a foot, or, say, $\frac{1}{83}$ of an inch, per annum.

Well, I do not know that any one is prepared to maintain that, even making all needful allowances, the stratified rocks may not have been formed, on the average, at the rate of $\frac{1}{83}$ of an inch per annum. I suppose that if such could be shown to be the limit of world-growth, we could put up with the allowance without feeling that our speculations had undergone any revolution. And perhaps, after all, the qualifying phrase " some such period " may not necessitate the assumption of more than $\frac{1}{166}$, or $\frac{1}{249}$, or $\frac{1}{332}$ of an inch of deposit per year, which, of course, would give us still more ease and comfort.

But, it may be said, that it is biology, and not geology, which asks for so much time—that the succession of life demands vast intervals ; but this appears to me to be reasoning in a circle. Biology takes her time from geology. The only reason we have for believing in the slow rate of the change in living forms is the fact that they persist through a series of deposits which, geology informs us, have taken a long while to make. If the geological clock is wrong, all the naturalist will have to do is to modify his notions of the rapidity of change accordingly. And I venture to point out that, when we are told that the limitation of the period during which living beings have inhabited this planet to one, two, or three hundred million years requires a complete revolution in geological speculation, the *onus probandi* rests on the maker of the assertion, who brings forward not a shadow of evidence in its support.

Thus, if we accept the limitation of time placed before us by Sir W. Thomson, it is not obvious, on the face of the matter, that we shall have to alter, or reform, our ways in any appreciable degree ; and we may therefore proceed with much calmness, and indeed much indifference, as to the result, to inquire whether that limitation is justified by the arguments employed in its support.

.These arguments are three in number :—

I. The first is based upon the undoubted fact that the tides tend to retard the rate of the earth's rotation upon its axis. That this must be so is obvious, if one considers, roughly, that the tides result from the pull which the sun and the moon exert upon the sea, causing it to act as a sort of break upon the rotating solid earth.

Kant, who was by no means a mere "abstract philosopher," but a good mathematician and well versed in the physical science of his time, not only proved this in an essay of exquisite clearness and intelligibility, now more than a century old,[1] but deduced from it some of its more important consequences, such as the constant turning of one face of the moon towards the earth.

But there is a long step from the demonstration of a tendency to the estimation of the practical value of that

---

[1] "Untersuchung der Frage ob die Erde in ihrer Umdrehung um die Achse, wodurch sie die Abwechselung des Tages und der Nacht hervorbringt, einige Veränderung seit den ersten Zeiten ihres Ursprunges erlitten habe, &c."—KANT'S *Sämmtliche Werke*, Bd. i. p. 178.

tendency, which is all with which we are at present concerned. The facts bearing on this point appear to stand as follow :—

It is a matter of observation that the moon's mean motion is (and has for the last 3,000 years been) undergoing an acceleration, relatively to the rotation of the earth. Of course this may result from one of two causes : the moon may really have been moving more swiftly in its orbit ; or the earth may have been rotating more slowly on its axis.

Laplace believed he had accounted for this phænomenon by the fact that the eccentricity of the earth's orbit has been diminishing throughout these 3,000 years. This would produce a diminution of the mean attraction of the sun on the moon ; or, in other words, an increase in the attraction of the earth on the moon ; and, consequently, an increase in the rapidity of the orbital motion of the latter body. Laplace, therefore, laid the responsibility of the acceleration upon the moon ; and if his views were correct, the tidal retardation must either be insignificant in amount, or be counteracted by some other agency.

Our great astronomer, Adams, however, appears to have found a flaw in Laplace's calculation, and to have shown that only half the observed retardation could be accounted for in the way he had suggested. There remains, therefore, the other half to be accounted for ; and here, in the absence of all positive knowledge, three sets of hypotheses have been suggested.

(*a.*) M. Delaunay suggests that the earth is at fault, in consequence of the tidal retardation. Messrs. Adams, Thomson, and Tait work out this suggestion, and, " on

a certain assumption as to the proportion of retardations
due to the sun and the moon," find the earth may lose
twenty-two seconds of time in a century from this cause.[1]

(*b.*) But M. Dufour suggests that the retardation of the
earth (which is hypothetically assumed to exist) may be
due in part, or wholly, to the increase of the moment
of inertia of the earth by meteors falling upon its surface.
This suggestion also meets with the entire approval of
Sir W. Thomson, who shows that meteor-dust, accumu-
lating at the rate of one foot in 4,000 years, would
account for the remainder of retardation.[2]

(*c.*) Thirdly, Sir W. Thomson brings forward an hypothesis of his own with respect to the cause of the hypothetical retardation of the earth's rotation :—

" Let us suppose ice to melt from the polar regions
(20° round each pole, we may say) to the extent of
something more than a foot thick, enough to give 1·1
foot of water over those areas, or 0·006 of a foot of
water if spread over the whole globe, which would, in
reality, raise the sea-level by only some such undiscover-
able difference as three-fourths of an inch or an inch.
This, or the reverse, which we believe might happen any
year, and could certainly not be detected without far
more accurate observations and calculations for the mean
sea-level than any hitherto made, would slacken or
quicken the earth's rate as a timekeeper by one-tenth of
a second per year."[3]

I do not presume to throw the slightest doubt upon
the accuracy of any of the calculations made by such
distinguished mathematicians as those who have made
the suggestions I have cited. On the contrary, it is

[1] Sir W. Thomson, loc. cit., p. 14.　　　[2] Loc. cit., p. 27.　　　[3] Ibid.

necessary to my argument to assume that they are all correct. But I desire to point out that this seems to be one of the many cases in which the admitted accuracy of mathematical processes is allowed to throw a wholly inadmissible appearance of authority over the results obtained by them. Mathematics may be compared to a mill of exquisite workmanship, which grinds you stuff of any degree of fineness ; but, nevertheless, what you get out depends on what you put in; and as the grandest mill in the world will not extract wheat-flour from peascods, so pages of formulæ will not get a definite result out of loose data.

In the present instance it appears to be admitted :—

1. That it is not absolutely certain, after all, whether the moon's mean motion is undergoing acceleration, or the earth's rotation retardation.[1] And yet this is the key of the whole position.

2. If the rapidity of the earth's rotation is diminishing, it is not certain how much of that retardation is due to tidal friction,—how much to meteors,—how much to possible excess of melting over accumulation of polar ice, during the period covered by observation, which amounts, at the outside, to not more than 2,600 years.

3. The effect of a different distribution of land and water in modifying the retardation caused by tidal friction, and of reducing it, under some circumstances, to a minimum, does not appear to be taken into account.

4. During the Miocene epoch the polar ice was certainly many feet thinner than it has been during, or

---

[1] It will be understood that I do not wish to deny that the earth's rotation *may be* undergoing retardation.

since, the Glacial epoch. Sir W. Thomson tells us that
the accumulation of something more than a foot of
ice around the poles (which implies the withdrawal of,
say, an inch of water from the general surface of the
sea) will cause the earth to rotate quicker by one-tenth
of a second per annum. It would appear, therefore,
that the earth may have been rotating, throughout the
whole period which has elapsed from the commencement
of the Glacial epoch down to the present time, one, or
more, seconds per annum quicker than it rotated during
the Miocene epoch.

But, according to Sir W. Thomson's calculation, tidal
retardation will only account for a retardation of 22″ in
a century, or $\frac{22}{100}$ (say $\frac{1}{5}$) of a second per annum.

Thus, assuming that the accumulation of polar ice
since the Miocene epoch has only been sufficient to
produce ten times the effect of a coat of ice one foot
thick, we shall have an accelerating cause which covers
all the loss from tidal action, and leaves a balance
of $\frac{4}{5}$ a second per annum in the way of acceleration.

If tidal retardation can be thus checked and over-
thrown by other temporary conditions, what becomes
of the confident assertion, based upon the assumed uniformity of tidal retardation, that ten thousand million
years ago the earth must have been rotating more than
twice as fast as at present, and, therefore, that we
geologists are "in direct opposition to the principles
of Natural Philosophy" if we spread geological history
over that time ?

II. The second argument is thus stated by Sir W.
Thomson :—"An article, by myself, published in 'Macmillan's Magazine' for March 1862, on the age of the

sun's heat, explains results of investigation into various questions as to possibilities regarding the amount of heat that the sun could have, dealing with it as you would with a stone, or a piece of matter, only taking into account the sun's dimensions, which showed it to be possible that the sun may have already illuminated the earth for as many as one hundred million years, but at the same time rendered it almost certain that he had not illuminated the earth for five hundred millions of years. The estimates here are necessarily very vague ; but yet, vague as they are, I do not know that it is possible, upon any reasonable estimate founded on known properties of matter, to say that we can believe the sun has really illuminated the earth for five hundred million years."[1]

I do not wish to " Hansardize " Sir William Thomson by laying much stress on the fact that, only fifteen years ago, he entertained a totally different view of the origin of the sun's heat, and believed that the energy radiated from year to year was supplied from year to year— a doctrine which would have suited Hutton perfectly. But the fact that so eminent a physical philosopher has, thus recently, held views opposite to those which he now entertains, and that he confesses his own estimates to be " very vague," justly entitles us to disregard those estimates, if any distinct facts on our side go against them. However, I am not aware that such facts exist. As I have already said, for anything I know, one, two, or three hundred millions of years may serve the needs of geologists perfectly well.

III. The third line of argument is based upon the

[1] Loc. cit., p. 20.

temperature of the interior of the earth.   Sir W. Thomson refers to certain investigations which prove that the present thermal condition of the interior of the earth implies either a heating of the earth within the last 20,000 years of as much as 100° F., or a greater heating all over the surface at some time further back than 20,000 years, and then proceeds thus :—

" Now, are geologists prepared to admit that, at some time within the last 20,000 years, there has been all over the earth so high a temperature as that ?   I presume not; no geologist—no *modern* geologist—would for a moment admit the hypothesis that the present state of underground heat is due to a heating of the surface at so late a period as 20,000 years ago.   If that is not admitted, we are driven to a greater heat at some time more than 20,000 years ago.   A greater heating all over the surface than 100° Fahrenheit would kill nearly all existing plants and animals, I may safely say. Are modern geologists prepared to say that all life was killed off the earth 50,000, 100,000, or 200,000 years ago ?   For the uniformity theory, the further back the time of high surface-temperature is put the better ; but the further back the time of heating, the hotter it must have been.   The best for those who draw most largely on time is that which puts it furthest back ; and that is the theory that the heating was enough to melt the whole.   But even if it was enough to melt the whole, we must still admit some limit, such as fifty million years, one hundred million years, or two or three hundred million years ago.   Beyond that we cannot go." [1]

[1] Loc. cit., p. 24.

It will be observed that the "limit" is once again of the vaguest, ranging from 50,000,000 years to 300,000,000. And the reply is, once more, that, for anything that can be proved to the contrary, one or two hundred million years might serve the purpose, even of a thorough-going Huttonian uniformitarian, very well.

But if, on the other hand, the 100,000,000 or 200,000,000 years appear to be insufficient for geological purposes, we must closely criticise the method by which the limit is reached. The argument is simple enough. *Assuming* the earth to be nothing but a cooling mass, the quantity of heat lost per year, *supposing* the rate of cooling to have been uniform, multiplied by any given number of years, will be given the minimum temperature that number of years ago.

But is the earth nothing but a cooling mass, "like a hot-water jar such as is used in carriages," or "a globe of sandstone?" and has its cooling been uniform? An affirmative answer to both these questions seems to be necessary to the validity of the calculations on which Sir W. Thomson lays so much stress.

Nevertheless it surely may be urged that such affirmative answers are purely hypothetical, and that other suppositions have an equal right to consideration.

For example, is it not possible that, at the prodigious temperature which would seem to exist at 100 miles below the surface, all the metallic bases may behave as mercury does at a red heat, when it refuses to combine with oxygen; while, nearer the surface, and therefore at a lower temperature, they may enter into combination (as mercury does with oxygen a few degrees below its boiling-point) and so give rise to a heat totally distinct from that which they possess as cooling bodies? And has it not also been proved by recent researches that the quality of the atmosphere may immensely affect its permeability to heat; and, consequently, profoundly modify the rate of cooling the globe as a whole?

I do not think it can be denied that such conditions may exist, and may so greatly affect the supply, and the loss, of terrestrial heat as to destroy the value of any calculations which leave them out of sight.

My functions as your advocate are at an end. I speak with more than the sincerity of a mere advocate when I express the belief that the case against us has entirely broken down. The cry for reform which has been raised without, is superfluous, inasmuch as we have long been reforming from within, with all needful speed. And the critical examination of the grounds upon which the very grave charge of opposition to the principles of Natural Philosophy has been brought against us, rather shows that we have exercised a wise discrimination in declining, for the present, to meddle with our foundations.

XII.

THE ORIGIN OF SPECIES.

MR. DARWIN'S long-standing and well-earned scientific eminence probably renders him indifferent to that social notoriety which passes by the name of success; but if the calm spirit of the philosopher have not yet wholly superseded the ambition and the vanity of the carnal man within him, he must be well satisfied with the results of his venture in publishing the "Origin of Species." Overflowing the narrow bounds of purely scientific circles, the "species question" divides with Italy and the Volunteers the attention of general society. Everybody has read Mr. Darwin's book, or, at least, has given an opinion upon its merits or demerits; pietists, whether lay or ecclesiastic, decry it with the mild railing which sounds so charitable; bigots denounce it with ignorant invective; old ladies, of both sexes, consider it a decidedly dangerous book, and even savans, who have no better mud to throw, quote antiquated writers to show that its author is no better than an ape himself; while every philosophical thinker hails it as a veritable Whitworth gun in the armory of

liberalism ; and all competent naturalists and physiologists, whatever their opinions as to the ultimate fate
of the doctrines put forth, acknowledge that the work in
which they are embodied is a solid contribution to knowledge and inaugurates a new epoch in natural history.

Nor has the discussion of the subject been restrained
within the limits of conversation. When the public is
eager and interested, reviewers must minister to its
wants ; and the genuine *littérateur* is too much in the
habit of acquiring his knowledge from the book he
judges—as the Abyssinian is said to provide himself
with steaks from the ox which carries him—to be with-
held from criticism of a profound scientific work by the
mere want of the requisite preliminary scientific acquirement ; while, on the other hand, the men of science who
wish well to the new views, no less than those who
dispute their validity, have naturally sought opportunities of expressing their opinions. Hence it is not
surprising that almost all the critical journals have
noticed Mr. Darwin's work at greater or less length ;
and so many disquisitions, of every degree of excellence,
from the poor product of ignorance, too often stimulated
by prejudice, to the fair and thoughtful essay of the
candid student of Nature, have appeared, that it seems
an almost hopeless task to attempt to say anything new
upon the question.

But it may be doubted if the knowledge and acumen
of prejudged scientific opponents, or the subtlety of
orthodox special pleaders, have yet exerted their full
force in mystifying the real issues of the great controversy which has been set afoot, and whose end is hardly
likely to be seen by this generation ; so that, at this

eleventh hour, and even failing anything new, it may be useful to state afresh that which is true, and to put the fundamental positions advocated by Mr. Darwin in such a form that they may be grasped by those whose special studies lie in other directions. And the adoption of this course may be the more advisable, because notwithstanding its great deserts, and indeed partly on account of them, the "Origin of Species" is by no means an easy book to read—if by reading is implied the full comprehension of an author's meaning.

We do not speak jestingly in saying that it is Mr. Darwin's misfortune to know more about the question he has taken up than any man living. Personally and practically exercised in zoology, in minute anatomy, in geology ; a student of geographical distribution, not on maps and in museums only, but by long voyages and laborious collection ; having largely advanced each of these branches of science, and having spent many years in gathering and sifting materials for his present work, the store of accurately registered facts upon which the author of the "Origin of Species" is able to draw at will is prodigious.

But this very superabundance of matter must have been embarrassing to a writer who, for the present, can only put forward an abstract of his views ; and thence it arises, perhaps, that notwithstanding the clearness of the style, those who attempt fairly to digest the book find much of it a sort of intellectual pemmican—a mass of facts crushed and pounded into shape, rather than held together by the ordinary medium of an obvious logical bond : due attention will, without doubt, discover this bond, but it is often hard to find.

Again, from sheer want of room, much has to be taken for granted which might readily enough be proved; and hence, while the adept, who can supply the missing links in the evidence from his own knowledge, discovers fresh proof of the singular thoroughness with which all difficulties have been considered and all unjustifiable suppositions avoided, at every reperusal of Mr. Darwin's pregnant paragraphs, the novice in biology is apt to complain of the frequency of what he fancies is gratuitous assumption.

Thus while it may be doubted if, for some years, any one is likely to be competent to pronounce judgment on all the issues raised by Mr. Darwin, there is assuredly abundant room for him, who, assuming the humbler, though perhaps as useful, office of an interpreter between the "Origin of Species" and the public, contents himself with endeavouring to point out the nature of the problems which it discusses; to distinguish between the ascertained facts and the theoretical views which it contains; and finally, to show the extent to which the explanation it offers satisfies the requirements of scientific logic. At any rate, it is this office which we propose to undertake in the following pages.

It may be safely assumed that our readers have a general conception of the nature of the objects to which the word "species" is applied; but it has, perhaps, occurred to few, even of those who are naturalists *ex professo*, to reflect, that, as commonly employed, the term has a double sense and denotes two very different orders of relations. When we call a group of animals, or of plants, a species, we may imply thereby either, that all these animals or plants have some common peculiarity of form or structure ; or, we may mean that they possess some common functional character. That part of biological science which deals with form and structure is called Morphology—that which concerns itself with function, Physiology—so that we may conveniently speak of these two senses, or aspects, of " species "—the one as morphological, the other as physiological. Regarded from the former point of view, a species is nothing more than a kind of animal or plant, which is distinctly definable from all others, by certain constant, and not merely sexual, morphological peculiarities. Thus horses form a species, because the group of animals to which that name is applied is distinguished from all others in the world by the following constantly associated characters. They have 1. A vertebral column ; 2. Mammæ ; 3. A placental embryo ; 4. Four legs ; 5. A single well-developed toe in each foot provided with a hoof ; 6. A bushy tail ; and 7. Callosities on the inner sides of both the fore and the hind legs. The asses, again, form a distinct species, because, with the same characters, as far as the fifth in the above list, all asses have tufted tails, and have callosities only on the inner side of the fore legs. If animals were discovered having the general characters of the horse, but sometimes with callosities only on the fore legs, and more or less tufted tails ; or animals having the general characters of the ass, but with more or less bushy tails, and sometimes with callosities on both pairs of legs, besides being intermediate in other respects—the two species would have to be merged into one. They could no longer be regarded as morphologically distinct species, for they would not be distinctly definable one from the other.

However bare and simple this definition of species may appear to be, we confidently appeal to all practical naturalists, whether zoologists, botanists, or palæontologists, to say if, in the vast majority of cases, they know, or mean to affirm, anything more of the group of animals or plants they so denominate than what has just been stated. Even the most decided advocates of the received doctrines respecting species admit this.

"I apprehend," says Professor Owen,[1] "that few naturalists now-a-days, in describing and proposing a name for what they call 'a new *species*,' use that term to signify what was meant by it twenty or thirty years ago; that is, an originally distinct creation, maintaining its primitive distinction by obstructive generative peculiarities. The proposer of the new species now intends to state no more than he actually knows; as, for example, that the differences on which he founds the specific character are constant in individuals of both sexes, so far as observation has reached; and that they are not due to domestication or to artificially superinduced external circumstances, or to any outward influence within his cognizance; that the species is wild, or is such as it appears by Nature."

If we consider, in fact, that by far the largest proportion of recorded existing species are known only by the study of their skins, or bones, or other lifeless exuvia; that we are acquainted with none, or next to none, of their physiological peculiarities, beyond those which can be deduced from their structure, or are open to cursory observation; and that we cannot hope to learn more of any of those extinct forms of life which now constitute no inconsiderable proportion of the known Flora and Fauna of the world : it is obvious that the definitions of these species can be only of a purely

[1] On the Osteology of the Chimpanzees and Orangs : Transactions of the Zoological Society, 1858.

structural or morphological character. It is probable that naturalists would have avoided much confusion of ideas if they had more frequently borne these necessary limitations of our knowledge in mind. But while it may safely be admitted that we are acquainted with only the morphological characters of the vast majority of species—the functional, or physiological, peculiarities of a few have been carefully investigated, and the result of that study forms a large and most interesting portion of the physiology of reproduction.

The student of Nature wonders the more and is astonished the less, the more conversant he becomes with her operations; but of all the perennial miracles she offers to his inspection, perhaps the most worthy of admiration is the development of a plant or of an animal from its embryo. Examine the recently laid egg of some common animal, such as a salamander or a newt. It is a minute spheroid in which the best microscope will reveal nothing but a structureless sac, enclosing a glairy fluid, holding granules in suspension. But strange possibilities lie dormant in that semi-fluid globule. Let a moderate supply of warmth reach its watery cradle, and the plastic matter undergoes changes so rapid and yet so steady and purposelike in their succession, that one can only compare them to those operated by a skilled modeller upon a formless lump of clay. As with an invisible trowel, the mass is divided and sub-divided into smaller and smaller portions, until it is reduced to an aggregation of granules not too large to build withal the finest fabrics of the nascent organism. And, then, it is as if a delicate finger traced out the line to be occupied by the spinal column, and moulded the

contour of the body ; pinching up the head at one end,
the tail at the other, and fashioning flank and limb into
due salamandrine proportions, in so artistic a way, that,
after watching the process hour by hour, one is almost
involuntarily possessed by the notion, that some more
subtle aid to vision than an achromatic, would show the
hidden artist, with his plan before him, striving with
skilful manipulation to perfect his work.

As life advances, and the young amphibian ranges the
waters, the terror of his insect contemporaries, not only
are the nutritious particles supplied by its prey, by the
addition of which to its frame growth takes place, laid
down, each in its proper spot, and in such due proportion
to the rest, as to reproduce the form, the colour and the
size, characteristic of the parental stock ; but even the
wonderful powers of reproducing lost parts possessed by
these animals are controlled by the same governing
tendency. Cut off the legs, the tail, the jaws, separately
or all together, and, as Spallanzani showed long ago,
these parts not only grow again, but the redintegrated
limb is formed on the same type as those which were
lost. The new jaw, or leg, is a newt's, and never by any
accident more like that of a frog. What is true of the
newt is true of every animal and of every plant ; the
acorn tends to build itself up again into a woodland
giant such as that from whose twig it fell ; the spore of
the humblest lichen reproduces the green or brown
incrustation which gave it birth ; and at the other end
of the scale of life, the child that resembled neither the
paternal nor the maternal side of the house would be
regarded as a kind of monster.

So that the one end to which, in all living beings, the

formative impulse is tending—the one scheme which the Archæus of the old speculators strives to carry out, seems to be to mould the offspring into the likeness of the parent. It is the first great law of reproduction, that the offspring tends to resemble its parent or parents, more closely than anything else.

Science will some day show us how this law is a necessary consequence of the more general laws which govern matter; but for the present, more can hardly be said than that it appears to be in harmony with them. We know that the phænomena of vitality are not something apart from other physical phænomena, but one with them; and matter and force are the two names of the one artist who fashions the living as well as the lifeless. Hence living bodies should obey the same great laws as other matter—nor, throughout Nature, is there a law of wider application than this, that a body impelled by two forces takes the direction of their resultant. But living bodies may be regarded as nothing but extremely complex bundles of forces held in a mass of matter, as the complex forces of a magnet are held in the steel by its coercive force; and, since the differences of sex are comparatively slight, or, in other words, the sum of the forces in each has a very similar tendency, their resultant, the offspring, may reasonably be expected to deviate but little from a course parallel to either, or to both.

Represent the reason of the law to ourselves by what physical metaphor or analogy we will, however, the great matter is to apprehend its existence and the importance of the consequences deducible from it. For things which are like to the same are like to one another, and if, in a great series of generations, every offspring is like its parent, it follows that all the offspring and all the parents must be like one another ; and that, given an original parental stock, with the opportunity of undisturbed multiplication, the law in question necessitates the production, in course of time, of an indefinitely large group, the whole of whose members are at once very similar and are blood relations, having descended from the same parent, or pair of parents. The proof that all the members of any given group of animals, or plants, had thus descended, would be ordinarily considered sufficient to entitle them to the rank of physiological species, for most physiologists consider species to be definable as " the offspring of a single primitive stock."

But though it is quite true that all those groups we call species *may*, according to the known laws of reproduction, have descended from a single stock, and though it is very likely they really have done so, yet this conclusion rests on deduction and can hardly hope to establish itself upon a basis of observation. And the primitiveness of the supposed single stock, which, after all, is the essential part of the matter, is not only a hypothesis, but one which has not a shadow of foundation, if by " primitive " be meant " independent of any other living being." A scientific definition, of which an unwarrantable hypothesis forms an essential part, carries its condemnation within itself; but even supposing such a definition were, in form, tenable, the physiologist who should attempt to apply it in Nature would soon find himself involved in great, if not inextricable difficulties. As we have said, it is indubitable that offspring *tend* to resemble the parental organism, but it is equally true that the similarity attained never amounts to identity, either in form or in structure. There is always a certain amount of deviation, not only from the precise characters of a single parent, but when, as in most animals and many plants, the sexes are lodged in distinct individuals, from an exact mean between the two parents. And indeed, on general principles, this slight deviation seems as intelligible as the general similarity, if we reflect how complex the co-operating " bundles of forces" are, and how improbable it is that, in any case, their true resultant shall coincide with any mean between the more obvious characters of the two parents. Whatever be its cause, however, the co-existence of this tendency to minor variation with the tendency to general similarity, is of vast importance in its bearing on the question of the origin of species.

As a general rule, the extent to which an offspring differs from its parent is slight enough ; but, occasionally, the amount of difference is much more strongly marked, and then the divergent offspring receives the name of a Variety. Multitudes, of what there is every reason to believe are such varieties, are known, but the origin of very few has been accurately recorded, and of these we will select two as more especially illustrative of the main features of variation. The first of them is that of the " Ancon," or " Otter" sheep, of which a careful account is given by Colonel David Humphreys, F.R.S., in a letter to Sir Joseph Banks, published in the Philosophical Transactions for 1813. It appears that one Seth Wright, the proprietor of a farm on the banks of the Charles River, in Massachusetts, possessed a flock of fifteen ewes and a ram of the ordinary kind. In the year 1791, one of the ewes presented her owner with a male lamb, differing, for no assignable reason, from its parents by a proportionally long body and short bandy legs, whence it was unable to emulate its relatives in those sportive leaps over the neighbours' fences, in which they were in the habit of indulging, much to the good farmer's vexation.

The second case is that detailed by a no less unexceptionable authority than Réaumur, in his " Art de faire éclore les Poulets." A Maltese couple, named Kelleia, whose hands and feet were constructed upon the ordinary human model, had born to them a son, Gratio, who possessed six perfectly moveable fingers on each hand and six toes, not quite so well formed, on each foot. No cause could be assigned for the appearance of this unusual variety of the human species.

Two circumstances are well worthy of remark in both these cases. In each, the variety appears to have arisen in full force, and, as it were, *per saltum ;* a wide and definite difference appearing, at once, between the Ancon ram and the ordinary sheep ; between the six-fingered and six-toed Gratio Kelleia and ordinary men. In neither case is it possible to point out any obvious reason for the appearance of the variety. Doubtless there were determining causes for these as for all other phænomena ; but they do not appear, and we can be tolerably certain that what are ordinarily understood as changes in physical conditions, as in climate, in food, or the like, did not take place and had nothing to do with the matter. It was no case of what is commonly called adaptation to circumstances ; but, to use a conveniently erroneous phrase, the variations arose spontaneously.

The fruitless search after final causes leads their pursuers a long way; but even those hardy teleologists, who are ready to break through all the laws of physics in chase of their favourite will-o'-the-wisp, may be puzzled to discover what purpose could be attained by the stunted legs of Seth Wright's ram or the hexadactyle members of Gratio Kelleia.

Varieties then arise we know not why; and it is more than probable that the majority of varieties have arisen in this "spontaneous" manner, though we are, of course, far from denying that they may be traced, in some cases, to distinct external influences; which are assuredly competent to alter the character of the tegumentary covering, to change colour, to increase or diminish the size of muscles, to modify constitution, and, among plants, to give rise to the metamorphosis of stamens into petals, and so forth. But however they may have arisen, what especially interests us at present is, to remark that, once in existence, varieties obey the fundamental law of reproduction that like tends to produce like, and their offspring exemplify it by tending to exhibit the same deviation from the parental stock as themselves. Indeed, there seems to be, in many instances, a pre-potent influence about a newly-arisen variety which gives it what one may call an unfair advantage over the normal descendants from the same stock. This is strikingly exemplified by the case of Gratio Kelleia, who married a woman with the ordinary pentadactyle extremities, and had by her four children, Salvator, George, André, and Marie. Of these children Salvator, the eldest boy, had six fingers and six toes, like his father; the second and third, also boys, had five fingers and five toes, like their mother, though the hands and feet of George were slightly deformed ; the last, a girl, had five fingers and five toes, but the thumbs were slightly deformed. The variety thus reproduced itself purely in the eldest, while the normal type reproduced itself purely in the third, and almost purely in the second and last : so that it would seem, at first, as if the normal type were more powerful than the variety. But all these children grew up and intermarried with normal wives and husband, and then, note what took place : Salvator had four children, three of whom exhibited the hexadactyle members of their grandfather and father, while the youngest had the pentadactyle limbs of the mother and grandmother ; so that here, notwithstanding a double pentadactyle dilution of the blood, the hexadactyle variety had the best of it. The same pre-potency of the variety was still more markedly exemplified in the progeny of two of the other children, Marie and George. Marie (whose thumbs only were deformed) gave birth to a boy with six toes, and three other normally formed children ; but George, who was not quite so pure a pentadactyle, begot, first, two girls, each of whom had six fingers and toes ; then a girl with six fingers on each hand and six toes on the right foot, but only five toes on the left ; and lastly, a boy with only five fingers and toes. In these instances, therefore, the variety, as it were, leaped over one generation to reproduce itself in full force in the next. Finally, the purely pentadactyle André was the father of many children, not one of whom departed from the normal parental type.

If a variation which approaches the nature of a monstrosity can strive thus forcibly to reproduce itself, it is not wonderful that less aberrant modifications should tend to be preserved even more strongly; and the history of the Ancon sheep is, in this respect, particularly instructive. With the " 'cuteness" characteristic of their nation, the neighbours of the Massachusetts farmer imagined it would be an excellent thing if all his sheep were imbued with the stay-at-home tendencies enforced by Nature upon the newly-arrived ram; and they advised Wright to kill the old patriarch of his fold, and install the Ancon ram in his place. The result justified their sagacious anticipations, and coincided very nearly with what occurred to the progeny of Gratio Kelleia. The young lambs were almost always either pure Ancons, or pure ordinary sheep.[1] But when sufficient Ancon sheep were obtained to interbreed with one another, it was found that the offspring was always pure Ancon. Colonel Humphreys, in fact, states that he was acquainted with only "one questionable case of a contrary nature." Here, then, is a remarkable and well-established instance, not only of a very distinct race being established *per saltum*, but of that race breeding "true" at once, and showing no mixed forms, even when crossed with another breed.

---

[1] Colonel Humphreys' statements are exceedingly explicit on this point :— "When an Ancon ewe is impregnated by a common ram, the increase resembles wholly either the ewe or the ram. The increase of the common ewe impregnated by an Ancon ram follows entirely the one or the other, without blending any of the distinguishing and essential peculiarities of both. Frequent instances have happened where common ewes have had twins by Ancon rams, when one exhibited the complete marks and features of the ewe, the other of the ram. The contrast has been rendered singularly striking, when one short-legged and one long-legged lamb, produced at a birth, have been seen sucking the dam at the same time."—*Philosophical Transactions*, 1813, Pt. I., pp. 89, 90.

By taking care to select Ancons of both sexes, for breeding from, it thus became easy to establish an extremely well-marked race; so peculiar that, even when herded with other sheep, it was noted that the Ancons kept together. And there is every reason to believe that the existence of this breed might have been indefinitely protracted; but the introduction of the Merino sheep, which were not only very superior to the Ancons in wool and meat, but quite as quiet and orderly, led to the complete neglect of the new breed, so that, in 1813, Colonel Humphreys found it difficult to obtain the specimen, whose skeleton was presented to Sir Joseph Banks. We believe that, for many years, no remnant of it has existed in the United States.

Gratio Kelleia was not the progenitor of a race of six-fingered men, as Seth Wright's ram became a nation of Ancon sheep, though the tendency of the variety to perpetuate itself appears to have been fully as strong, in the one case as in the other. And the reason of the difference is not far to seek. Seth Wright took care not to weaken the Ancon blood by matching his Ancon ewes with any but males of the same variety, while Gratio Kelleia's sons were too far removed from the patriarchal times to intermarry with their sisters; and his grandchildren seem not to have been attracted by their six-fingered cousins. In other words, in the one example a race was produced, because, for several generations, care was taken to *select* both parents of the breeding stock, from animals exhibiting a tendency to vary in the same direction; while, in the other, no race was evolved, because no such selection was exercised. A race is a propagated variety; and as, by the laws of reproduction, offspring tend to assume the parental form, they will be more likely to propagate a variation exhibited by both parents than that possessed by only one.

There is no organ of the body of an animal which may not, and does not, occasionally, vary more or less from the normal type; and there is no variation which may not be transmitted, and which, if selectively transmitted, may not become the foundation of a race. This great truth, sometimes forgotten by philosophers, has long been familiar to practical agriculturists and breeders: and upon it rest all the methods of improving the breeds of domestic animals, which, for the last century, have been followed with so much success in England. Colour, form, size, texture of hair or wool, proportions of various parts, strength or weakness of constitution, tendency to fatten or to remain lean, to give much or little milk, speed, strength, temper, intelligence, special instincts; there is not one of these characters whose transmission is not an every-day occurrence within the experience of cattle-breeders, stock-farmers, horse-dealers, and dog and poultry fanciers. Nay, it is only the other day that an eminent physiologist, Dr. Brown-Séquard, communicated to the Royal Society his discovery that epilepsy, artificially produced in guinea-pigs, by a means which he has discovered, is transmitted to their offspring.

But a race, once produced, is no more a fixed and immutable entity than the stock whence it sprang; variations arise among its members, and as these variations are transmitted like any others, new races may be developed out of the pre-existing ones *ad infinitum*, or, at least, within any limit at present determined. Given sufficient time and sufficiently careful selection, and the multitude of races which may arise from a common stock is as astonishing as are the extreme structural differences which they may present.   A remarkable example of this is to be found in the rock-pigeon, which Mr. Darwin has, in our opinion, satisfactorily demonstrated to be the progenitor of all our domestic pigeons, of which there are certainly more than a hundred well-marked races.   The most noteworthy of these races are, the four great stocks known to the "fancy" as tumblers, pouters, carriers, and fantails; birds which not only differ most singularly in size, colour, and habits, but in the form of the beak and of the skull: in the proportions of the beak to the skull ; in the number of tail-feathers ; in the absolute and relative size of the feet ; in the presence or absence of the uropygial gland ; in the number of vertebræ in the back ; in short, in precisely those characters in which the genera and species of birds differ from one another.

And it is most remarkable and instructive to observe, that none of these races can be shown to have been originated by the action of changes in what are commonly called external circumstances, upon the wild rock-pigeon.   On the contrary, from time immemorial, pigeon fanciers have had essentially similar methods of treating their pets, which have been housed, fed, protected and cared for in much the same way in all pigeonries.   In fact, there is no case better adapted than that of the pigeons, to refute the doctrine which one sees put forth on high authority, that "no other characters than those founded on the development of bone for the attachment of muscles" are capable of variation.   In precise contradiction of this hasty assertion, Mr. Darwin's researches prove that the skeleton of the wings in domestic pigeons has hardly varied at all from that of the wild type; while, on the other hand, it is in exactly those respects, such as the relative length of the beak and skull, the number of the vertebræ, and the number of the tail-feathers, in which muscular exertion can have no important influence, that the utmost amount of variation has taken place.

We have said that the following out of the properties exhibited by physiological species would lead us into difficulties, and at this point they begin to be obvious; for, if, as a result of spontaneous variation and of selective breeding, the progeny of a common stock may become separated into groups distinguished from one another by constant, not sexual, morphological characters, it is clear that the physiological definition of species is likely to clash with the morphological definition. No one would hesitate to describe the pouter and the tumbler as distinct species, if they were found fossil, or if their skins and skeletons were imported, as those of exotic wild birds commonly are—and, without doubt, if considered alone, they are good and distinct morphological species. On the other hand, they are not physiological species, for they are descended from a common stock, the rock-pigeon.

Under these circumstances, as it is admitted on all sides that races occur in Nature, how are we to know whether any apparently distinct animals are really of different physiological species, or not, seeing that the amount of morphological difference is no safe guide? Is there any test of a physiological species? The usual

answer of physiologists is in the affirmative. It is said
that such a test is to be found in the phænomena of
hybridization—in the results of crossing races, as compared with the results of crossing species.

So far as the evidence goes at present, individuals, of
what are certainly known to be mere races produced by
selection, however distinct they may appear to be, not
only breed freely together, but the offspring of such
crossed races are only perfectly fertile with one another.
Thus, the spaniel and the greyhound, the dray-horse and
the Arab, the pouter and the tumbler, breed together
with perfect freedom, and their mongrels, if matched
with other mongrels of the same kind, are equally fertile.

On the other hand, there can be no doubt that the
individuals of many natural species are either absolutely
infertile, if crossed with individuals of other species, or,
if they give rise to hybrid offspring, the hybrids so
produced are infertile when paired together. The horse
and the ass, for instance, if so crossed, give rise to the
mule, and there is no certain evidence of offspring ever
having been produced by a male and female mule. The
unions of the rock-pigeon and the ring-pigeon appear to
be equally barren of result. Here, then, says the physiologist, we have a means of distinguishing any two
true species from any two varieties. If a male and a
female, selected from each group, produce offspring, and
that offspring is fertile with others produced in the same
way, the groups are races and not species. If, on the
other hand, no result ensues, or if the offspring are
infertile with others produced in the same way, they are
true physiological species. The test would be an admirable one, if, in the first place, it were always practicable

to apply it, and if, in the second, it always yielded
results susceptible of a definite interpretation.   Unfortunately, in the great majority of cases, this touchstone
for species is wholly inapplicable.

The constitution of many wild animals is so altered
by confinement that they will not breed even with their
own females, so that the negative results obtained from
crosses are of no value ; and the antipathy of wild animals
of different species for one another, or even of wild and
tame members of the same species, is ordinarily so great,
that it is hopeless to look for such unions in Nature.
The hermaphrodism of most plants, the difficulty in the
way of ensuring the absence of their own, or the proper
working of other pollen, are obstacles of no less magnitude in applying the test to them.   And in both animals
and plants is superadded the further difficulty, that
experiments must be continued over a long time for the
purpose of ascertaining the fertility of the mongrel or
hybrid progeny, as well as of the first crosses from
which they spring.

Not only do these great practical difficulties lie in the
way of applying the hybridization test, but even when
this oracle can be questioned, its replies are sometimes
as doubtful as those of Delphi.   For example, cases are
cited by Mr. Darwin, of plants which are more fertile
with the pollen of another species than with their own ;
and there are others, such as certain *fuci*, whose male
element will fertilize the ovule of a plant of distinct species, while the males of the latter species are ineffective
with the females of the first.   So that, in the last-named
instance, a physiologist, who should cross the two species
in one way, would decide that they were true species ;

while another, who should cross them in the reverse
way, would, with equal justice, according to the rule,
pronounce them to be mere races.   Several plants, which
there is great reason to believe are mere varieties, are
almost sterile when crossed ; while both animals and
plants, which have always been regarded by naturalists
as of distinct species, turn out, when the test is applied,
to be perfectly fertile.   Again, the sterility or fertility
of crosses seems to bear no relation to the structural
resemblances or differences of the members of any two
groups.

Mr. Darwin has discussed this question with singular
ability and circumspection, and his conclusions are
summed up as follow, at page 276 of his work :—

"First crosses between forms sufficiently distinct to be ranked as
species, and their hybrids, are very generally, but not universally,
sterile.   The sterility is of all degrees, and is often so slight that the
two most careful experimentalists who have ever lived have come to
diametrically opposite conclusions in ranking forms by this test.   The
sterility is innately variable in individuals of the same species, and is
eminently susceptible of favourable and unfavourable conditions.   The
degree of sterility does not strictly follow systematic affinity, but is
governed by several curious and complex laws.   It is generally different, and sometimes widely different, in reciprocal crosses between
the same two species.   It is not always equal in degree in a first cross,
and in the hybrid produced from this cross.

"In the same manner as in grafting trees, the capacity of one
species or variety to take on another is incidental on generally
unknown differences in their vegetative systems ; so in crossing, the
greater or less facility of one species to unite with another is incidental on unknown differences in their reproductive systems.   There
is no more reason to think that species have been specially endowed
with various degrees of sterility to prevent them crossing and breeding
in Nature, than to think that trees have been specially endowed with
various and somewhat analogous degrees of difficulty in being grafted
together, in order to prevent them becoming inarched in our forests.

"The sterility of first crosses between pure species, which have their reproductive systems perfect, seems to depend on several circumstances; in some cases largely on the early death of the embryo. The sterility of hybrids which have their reproductive systems imperfect, and which have had this system and their whole organization disturbed by being compounded of two distinct species, seems closely allied to that sterility which so frequently affects pure species when their natural conditions of life have been disturbed. This view is supported by a parallelism of another kind; namely, that the crossing of forms, only slightly different, is favourable to the vigour and fertility of the offspring; and that slight changes in the conditions of life are apparently favourable to the vigour and fertility of all organic beings. It is not surprising that the degree of difficulty in uniting two species, and the degree of sterility of their hybrid offspring, should generally correspond, though due to distinct causes; for both depend on the amount of difference of some kind between the species which are crossed. Nor is it surprising that the facility of effecting a first cross, the fertility of hybrids produced from it, and the capacity of being grafted together—though this latter capacity evidently depends on widely different circumstances—should all run to a certain extent parallel with the systematic affinity of the forms which are subjected to experiment; for systematic affinity attempts to express all kinds of resemblance between all species.

"First crosses between forms known to be varieties, or sufficiently alike to be considered as varieties, and their mongrel offspring, are very generally, but not quite universally, fertile. Nor is this nearly general and perfect fertility surprising, when we remember how liable we are to argue in a circle with respect to varieties in a state of Nature; and when we remember that the greater number of varieties have been produced under domestication by the selection of mere external differences, and not of differences in the reproductive system. In all other respects, excluding fertility, there is a close general resemblance between hybrids and mongrels."—Pp. 276-8.

We fully agree with the general tenor of this weighty passage; but forcible as are these arguments, and little as the value of fertility or infertility as a test of species may be, it must not be forgotten that the really important fact, so far as the inquiry into the origin of species goes, is, that there are such things in Nature as groups

of animals and of plants, whose members are incapable
of fertile union with those of other groups ; and that
there are such things as hybrids, which are absolutely
sterile when crossed with other hybrids. For if such
phænomena as these were exhibited by only two of those
assemblages of living objects, to which the name of
species (whether it be used in its physiological or in
its morphological sense) is given, it would have to be
accounted for by any theory of the origin of species, and
every theory which could not account for it would be, so
far, imperfect.

Up to this point we have been dealing with matters of
fact, and the statements which we have laid before the
reader would, to the best of our knowledge, be admitted
to contain a fair exposition of what is at present known
respecting the essential properties of species, by all who
have studied the question. And whatever may be his theoretical views, no naturalist will probably be disposed to
demur to the following summary of that exposition :—

Living beings, whether animals or plants, are divisible
into multitudes of distinctly definable kinds, which are
morphological species. They are also divisible into
groups of individuals, which breed freely together, tending to reproduce their like, and are physiological species.
Normally resembling their parents, the offspring of
members of these species are still liable to vary, and the
variation may be perpetuated by selection, as a race,
which race, in many cases, presents all the characteristics
of a morphological species. But it is not as yet proved
that a race ever exhibits, when crossed with another race
of the same species, those phænomena of hybridization
which are exhibited by many species when crossed with

other species. On the other hand, not only is it not proved that all species give rise to hybrids infertile *inter se*, but there is much reason to believe that, in crossing, species exhibit every gradation from perfect sterility to perfect fertility.

Such are the most essential characteristics of species. Even were man not one of them—a member of the same system and subject to the same laws—the question of their origin, their causal connexion, that is, with the other phænomena of the universe, must have attracted his attention, as soon as his intelligence had raised itself above the level of his daily wants.

Indeed history relates that such was the case, and has embalmed for us the speculations upon the origin of living beings, which were among the earliest products of the dawning intellectual activity of man. In those early days positive knowledge was not to be had, but the craving after it needed, at all hazards, to be satisfied, and according to the country, or the turn of thought of the speculator, the suggestion that all living things arose from the mud of the Nile, from a primeval egg, or from some more anthropomorphic agency, afforded a sufficient resting-place for his curiosity. The myths of Paganism are as dead as Osiris or Zeus, and the man who should revive them, in opposition to the knowledge of our time, would be justly laughed to scorn; but the coeval imaginations current among the rude inhabitants of Palestine, recorded by writers whose very name and age are admitted by every scholar to be unknown, have unfortunately not yet shared their fate, but, even at this day, are regarded by nine-tenths of the civilized world as the

authoritative standard of fact and the criterion of the
justice of scientific conclusions, in all that relates to the
origin of things, and, among them, of species. In this
nineteenth century, as at the dawn of modern physical
science, the cosmogony of the semi-barbarous Hebrew is
the incubus of the philosopher and the opprobrium of the
orthodox. Who shall number the patient and earnest
seekers after truth, from the days of Galileo until now,
whose lives have been embittered and their good name
blasted by the mistaken zeal of Bibliolaters? Who
shall count the host of weaker men whose sense of
truth has been destroyed in the effort to harmonize
impossibilities—whose life has been wasted in the
attempt to force the generous new wine of Science
into the old bottles of Judaism, compelled by the outcry
of the same strong party?

It is true that if philosophers have suffered, their
cause has been amply avenged. Extinguished theolo-
gians lie about the cradle of every science as the
strangled snakes beside that of Hercules; and history
records that whenever science and orthodoxy have been
fairly opposed, the latter has been forced to retire from
the lists, bleeding and crushed, if not annihilated;
scotched, if not slain. But orthodoxy is the Bourbon
of the world of thought. It learns not, neither can it
forget; and though, at present, bewildered and afraid
to move, it is as willing as ever to insist that the first
chapter of Genesis contains the beginning and the end of
sound science; and to visit, with such petty thunderbolts
as its half-paralysed hands can hurl, those who refuse to
degrade Nature to the level of primitive Judaism.

Philosophers, on the other hand, have no such aggressive tendencies. With eyes fixed on the noble goal to which "per aspera et ardua" they tend, they may, now and then, be stirred to momentary wrath by the unnecessary obstacles with which the ignorant, or the malicious, encumber, if they cannot bar, the difficult path; but why should their souls be deeply vexed? The majesty of Fact is on their side, and the elemental forces of Nature are working for them. Not a star comes to the meridian at its calculated time but testifies to the justice of their methods—their beliefs are "one with the falling rain and with the growing corn." By doubt they are established, and open inquiry is their bosom friend. Such men have no fear of traditions however venerable, and no respect for them when they become mischievous and obstructive; but they have better than mere antiquarian business in hand, and if dogmas, which ought to be fossil but are not, are not forced upon their notice, they are too happy to treat them as non-existent.

The hypotheses respecting the origin of species which profess to stand upon a scientific basis, and, as such, alone demand serious attention, are of two kinds. The one, the "special creation" hypothesis, presumes every species to have originated from one or more stocks, these not being the result of the modification of any other form of living matter — or arising by natural agencies—but being produced, as such, by a supernatural creative act.

The other, the so-called "transmutation" hypothesis, considers that all existing species are the result of the modification of pre-existing species, and those of their predecessors, by agencies similar to those which at the present day produce varieties and races, and therefore in an altogether natural way; and it is a probable, though not a necessary consequence of this hypothesis, that all living beings have arisen from a single stock. With respect to the origin of this primitive stock, or stocks, the doctrine of the origin of species is obviously not necessarily concerned. The transmutation hypothesis, for example, is perfectly consistent either with the conception of a special creation of the primitive germ, or with the supposition of its having arisen, as a modification of inorganic matter, by natural causes.

The doctrine of special creation owes its existence very largely to the supposed necessity of making science accord with the Hebrew cosmogony; but it is curious to observe that, as the doctrine is at present maintained by men of science, it is as hopelessly inconsistent with the Hebrew view as any other hypothesis.

If there be any result which has come more clearly out of geological investigation than another, it is, that the vast series of extinct animals and plants is not divisible, as it was once supposed to be, into distinct groups, separated by sharply marked boundaries. There are no great gulfs between epochs and formations—no successive periods marked by the appearance of plants, of water animals, and of land animals, *en masse.* Every year adds to the list of links between what the older geologists supposed to be widely separated epochs: witness the crags linking the drift with the older tertiaries; the Maestricht beds linking the tertiaries with the chalk; the St. Cassian beds exhibiting an abundant fauna of mixed mesozoic and palæozoic types, in rocks of an epoch once supposed to be eminently poor in life;

witness, lastly, the incessant disputes as to whether a
given stratum shall be reckoned devonian or carboniferous, silurian or devonian, cambrian or silurian.

This truth is further illustrated in a most interesting
manner by the impartial and highly competent testimony
of M. Pictet, from whose calculations of what percentage
of the genera of animals, existing in any formation, lived
during the preceding formation, it results that in no
case is the proportion less than *one-third*, or 33 per
cent. It is the triassic formation, or the commencement
of the mesozoic epoch, which has received this smallest
inheritance from preceding ages. The other formations
not uncommonly exhibit 60, 80, or even 94 per cent.
of genera in common with those whose remains are
imbedded in their predecessor. Not only is this true,
but the subdivisions of each formation exhibit new
species characteristic of, and found only in, them ; and,
in many cases, as in the lias for example, the separate
beds of these subdivisions are distinguished by well-
marked and peculiar forms of life. A section, a hundred
feet thick, will exhibit, at different heights, a dozen
species of ammonite, none of which passes beyond its
particular zone of limestone, or clay, into the zone below
it or into that above it ; so that those who adopt the
doctrine of special creation must be prepared to admit,
that at intervals of time, corresponding with the thickness
of these beds, the Creator thought fit to interfere with
the natural course of events for the purpose of making
a new ammonite. It is not easy to transplant oneself
into the frame of mind of those who can accept such a
conclusion as this, on any evidence short of absolute
demonstration ; and it is difficult to see what is to be

gained by so doing, since, as we have said, it is obvious
that such a view of the origin of living beings is utterly
opposed to the Hebrew cosmogony. Deserving no aid
from the powerful arm of bibliolatry, then, does the
received form of the hypothesis of special creation derive
any support from science or sound logic ? Assuredly
not much. The arguments brought forward in its favour
all take one form : If species were not supernaturally
created, we cannot understand the facts x, or y, or z;
we cannot understand the structure of animals or plants,
unless we suppose they were contrived for special ends;
we cannot understand the structure of the eye, except
by supposing it to have been made to see with ; we
cannot understand instincts, unless we suppose animals
to have been miraculously endowed with them.

As a question of dialectics, it must be admitted that
this sort of reasoning is not very formidable to those
who are not to be frightened by consequences. It is an
argumentum ad ignorantiam—take this explanation or
be ignorant. But suppose we prefer to admit our ignorance rather than adopt a hypothesis at variance with
all the teachings of Nature ? Or, suppose for a moment
we admit the explanation, and then seriously ask ourselves how much the wiser are we ; what does the
explanation explain ? Is it any more than a grandilo-
quent way of announcing the fact, that we really know
nothing about the matter ? A phænomenon is explained
when it is shown to be a case of some general law of
Nature ; but the supernatural interposition of the Creator
can, by the nature of the case, exemplify no law, and if
species have really arisen in this way, it is absurd to
attempt to discuss their origin.

Or, lastly, let us ask ourselves whether any amount of evidence which the nature of our faculties permits us to attain, can justify us in asserting that any phæno-menon is out of the reach of natural causation. To this end it is obviously necessary that we should know all the consequences to which all possible combinations, continued through unlimited time, can give rise. If we knew these, and found none competent to originate species, we should have good ground for denying their origin by natural causation. Till we know them, any hypothesis is better than one which involves us in such miserable presumption.

But the hypothesis of special creation is not only a mere specious mask for our ignorance; its existence in Biology marks the youth and imperfection of the science. For what is the history of every science but the his-tory of the elimination of the notion of creative, or other interferences, with the natural order of the phæno-mena which are the subject-matter of that science? When Astronomy was young "the morning stars sang together for joy," and the planets were guided in their courses by celestial hands. Now, the harmony of the stars has resolved itself into gravitation according to the inverse squares of the distances, and the orbits of the planets are deducible from the laws of the forces which allow a schoolboy's stone to break a window. The lightning was the angel of the Lord; but it has pleased Providence, in these modern times, that science should make it the humble messenger of man, and we know that every flash that shimmers about the horizon on a summer's evening is determined by ascertainable conditions, and that its direction and brightness might, it our knowledge of these were great enough, have been calculated.

The solvency of great mercantile companies rests on the validity of the laws which have been ascertained to govern the seeming irregularity of that human life which the moralist bewails as the most uncertain of things ; plague, pestilence, and famine are admitted, by all but fools, to be the natural result of causes for the most part fully within human control, and not the unavoidable tortures inflicted by wrathful Omnipotence upon his helpless handiwork.

Harmonious order governing eternally continuous progress—the web and woof of matter and force interweaving by slow degrees, without a broken thread; that veil which lies between us and the Infinite—that universe which alone we know or can know ; such is the picture which science draws of the world, and in proportion as any part of that picture is in unison with the rest, so may we feel sure that it is rightly painted. Shall Biology alone remain out of harmony with her sister sciences ?

Such arguments against the hypothesis of the direct creation of species as these are plainly enough deducible from general considerations ; but there are, in addition, phænomena exhibited by species themselves, and yet not so much a part of their very essence as to have required earlier mention, which are in the highest degree perplexing, if we adopt the popularly accepted hypothesis. Such are the facts of distribution in space and in time ; the singular phænomena brought to light by the study of development ; the structural relations of species upon which our systems of classification are founded ; the great doctrines of philosophical anatomy, such as that of homology, or of the community of structural plan exhibited by large groups of species differing very widely in their habits and functions.

The species of animals which inhabit the sea on opposite sides of the isthmus of Panama are wholly distinct ;[1] the animals and plants which inhabit islands are commonly distinct from those of the neighbouring mainlands, and yet have a similarity of aspect. The mammals of the latest tertiary epoch in the Old and New Worlds belong to the same genera, or family groups, as those which now inhabit the same great geographical area. The crocodilian reptiles which existed in the earliest secondary epoch were similar in general structure to those now living, but exhibit slight differences in their vertebræ, nasal passages, and one or two other points. The guinea-pig has teeth which are shed before it is born, and hence can never subserve the masticatory purpose for which they seem contrived, and, in like manner, the female dugong has tusks which never cut the gum. All the members of the same great group run through similar conditions in their development, and all their parts, in the adult state, are arranged according to the same plan. Man is more like a gorilla than a gorilla is like a lemur. Such are a few, taken at random, among the multitudes of similar facts which modern research has established ; but when the student seeks for an explanation of them from the supporters of the received hypothesis of the origin of species, the reply he receives is, in substance, of Oriental sim-

---

[1] Recent investigations tend to show that this statement is not strictly accurate.—1870.

plicity and brevity—"Mashallah! it so pleases God!"
There are different species on opposite sides of the
isthmus of Panama, because they were created different
on the two sides. The pliocene mammals are like the
existing ones, because such was the plan of creation;
and we find rudimental organs and similarity of plan,
because it has pleased the Creator to set before himself
a "divine exemplar or archetype," and to copy it in his
works; and somewhat ill, those who hold this view
imply, in some of them. That such verbal hocus-pocus
should be received as science will one day be regarded
as evidence of the low state of intelligence in the nineteenth century, just as we amuse ourselves with the
phraseology about Nature's abhorrence of a vacuum,
wherewith Torricelli's compatriots were satisfied to
explain the rise of water in a pump. And be it recollected that this sort of satisfaction works not only
negative but positive ill, by discouraging inquiry, and
so depriving man of the usufruct of one of the most
fertile fields of his great patrimony, Nature.

The objections to the doctrine of the origin of species
by special creation which have been detailed, must have
occurred, with more or less force, to the mind of every
one who has seriously and independently considered the
subject. It is therefore no wonder that, from time to
time, this hypothesis should have been met by counter
hypotheses, all as well, and some better, founded than
itself; and it is curious to remark that the inventors of
the opposing views seem to have been led into them
as much by their knowledge of geology, as by their
acquaintance with biology. In fact, when the mind has
once admitted the conception of the gradual production

of the present physical state of our globe, by natural causes operating through long ages of time, it will be little disposed to allow that living beings have made their appearance in another way, and the speculations of De Maillet and his successors are the natural complement of Scilla's demonstration of the true nature of fossils.

A contemporary of Newton and of Leibnitz, sharing therefore in the intellectual activity of the remarkable age which witnessed the birth of modern physical science, Benoît de Maillet spent a long life as a consular agent of the French Government in various Mediterranean ports. For sixteen years, in fact, he held the office of Consul-General in Egypt, and the wonderful phænomena offered by the valley of the Nile appear to have strongly impressed his mind, to have directed his attention to all facts of a similar order which came within his observation, and to have led him to speculate on the origin of the present condition of our globe and of its inhabitants. But, with all his ardour for science, De Maillet seems to have hesitated to publish views which, notwithstanding the ingenious attempts to reconcile them with the Hebrew hypothesis contained in the preface to "Telliamed," were hardly likely to be received with favour by his contemporaries.

But a short time had elapsed since more than one of the great anatomists and physicists of the Italian school had paid dearly for their endeavours to dissipate some of the prevalent errors ; and their illustrious pupil, Harvey, the founder of modern physiology, had not fared so well, in a country less oppressed by the benumbing influences of theology, as to tempt any man to follow his example. Probably not uninfluenced by these considerations, his

Catholic majesty's Consul-General for Egypt kept his theories to himself throughout a long life, for "Telliamed," the only scientific work which is known to have proceeded from his pen, was not printed till 1735, when its author had reached the ripe age of seventy-nine ; and though De Maillet lived three years longer, his book was not given to the world before 1748. Even then it was anonymous to those who were not in the secret of the anagramatic character of its title ; and the preface and dedication are so worded as, in case of necessity, to give the printer a fair chance of falling back on the excuse that the work was intended for a mere *jeu d'esprit*

The speculations of the supposititious Indian sage, though quite as sound as those of many a " Mosaic Geology," which sells exceedingly well, have no great value if we consider them by the light of modern science. The waters are supposed to have originally covered the whole globe ; to have deposited the rocky masses which compose its mountains by processes comparable to those which are now forming mud, sand, and shingle ; and then to have gradually lowered their level, leaving the spoils of their animal and vegetable inhabitants embedded in the strata. As the dry land appeared, certain of the aquatic animals are supposed to have taken to it, and to have become gradually adapted to terrestrial and aërial modes of existence. But if we regard the general tenor and style of the reasoning in relation to the state of knowledge of the day, two circumstances appear very well worthy of remark. The first, that De Maillet had a notion of the modifiability of living forms (though without any precise information on the subject), and how such modifiability might account for the origin of species; the second, that he very clearly apprehended the great modern geological doctrine, so strongly insisted upon by Hutton, and so ably and comprehensively expounded by Lyell, that we must look to existing causes for the explanation of past geological events. Indeed, the following passage of the preface, in which De Maillet is supposed to speak of the Indian philosopher Telliamed, his *alter ego*, might have been written by the most philosophical uniformitarian of the present day :—

" Ce qu'il y a d'étonnant, est que pour arriver à ces connoissances i semble avoir perverti l'ordre naturel, puis-qu'au lieu de s'attacher d'abord à rechercher l'origine de notre globe il a commencé par travailler à s'instruire de la nature. Mais à l'entendre, ce renversement de l'ordre a été pour lui l'effet d'un génie favorable qui l'a conduit pas à pas et comme par la main aux découvertes les plus sublimes. C'est en décomposant la substance de ce globe par une anatomie exacte de toutes ses parties qu'il a premièrement appris de quelles matières il etait composé et quels arrangemens ces mêmes matières observaient entre elles. Ces lumières jointes à l'esprit de comparaison toujours nécessaire à quiconque entreprend de percer les voiles dont la nature aime à se cacher, ont servi de guide à notre philosophe pour parvenir à des connoissances plus intéressantes. Par la matière et l'arrangement de ces compositions il prétend avoir reconnu quelle est la véritable origine de ce globe que nous habitons, comment et par qui il a été formé."—Pp. xix. xx.

But De Maillet was before his age, and as could hardly fail to happen to one who speculated on a zoological and botanical question before Linnæus, and on a physiological problem before Haller, he fell into great errors here and there ; and hence, perhaps, the general neglect of his work. Robinet's speculations are rather behind, than in advance of, those of De Maillet ; and though Linnæus may have played with the hypothesis of transmutation, it obtained no serious support until

Lamarck adopted it, and advocated it with great ability in his "Philosophie Zoologique."

Impelled towards the hypothesis of the transmutation of species, partly by his general cosmological and geological views; partly by the conception of a graduated, though irregularly branching, scale of being, which had arisen out of his profound study of plants and of the lower forms of animal life, Lamarck, whose general line of thought often closely resembles that of De Maillet, made a great advance upon the crude and merely speculative manner in which that writer deals with the question of the origin of living beings, by endeavouring to find physical causes competent to effect that change of one species into another, which De Maillet had only supposed to occur. And Lamarck conceived that he had found in Nature such causes, amply sufficient for the purpose in view. It is a physiological fact, he says, that organs are increased in size by action, atrophied by inaction; it is another physiological fact that modifications produced are transmissible to offspring. Change the actions of an animal, therefore, and you will change its structure, by increasing the development of the parts newly brought into use and by the diminution of those less used; but by altering the circumstances which surround it you will alter its actions, and hence, in the long run, change of circumstance must produce change of organization. All the species of animals, therefore, are, in Lamarck's view, the result of the indirect action of changes of circumstance upon those primitive germs which he considered to have originally arisen, by spontaneous generation, within the waters of the globe. It is curious, however, that Lamarck should insist so strongly[1] as he has done, that circumstances never in any degree directly modify the form or the organization of animals, but only operate by changing their wants and consequently their actions; for he thereby brings upon himself the obvious question, how, then, do plants, which cannot be said to have wants or actions, become modified? To this he replies, that they are modified by the changes in their nutritive processes, which are effected by changing circumstances; and it does not seem to have occurred to him that such changes might be as well supposed to take place among animals.

When we have said that Lamarck felt that mere speculation was not the way to arrive at the origin of species, but that it was necessary, in order to the establishment of any sound theory on the subject, to discover by observation or otherwise, some *vera causa*, competent to give rise to them; that he affirmed the true order of classification to coincide with the order of their development one from another; that he insisted on the necessity of allowing sufficient time, very strongly; and that all the varieties of instinct and reason were traced back by him to the same cause as that which has given rise to species, we have enumerated his chief contributions to the advance of the question. On the other hand, from his ignorance of any power in Nature competent to modify the structure of animals, except the development of parts, or atrophy of them, in consequence of a change of needs, Lamarck was led to attach infinitely greater weight than it deserves to this agency, and the absurdities into which he was led have met with deserved condemnation. Of the struggle for existence, on which,

---

[1] See Phil. Zoologique, vol. i. p. 222, et seq.

as, we shall see, Mr. Darwin lays such great stress, he had
no conception; indeed, he doubts whether there really
are such things as extinct species, unless they be such
large animals as may have met their death at the hands
of man; and so little does he dream of there being any
other destructive causes at work, that, in discussing
the possible existence of fossil shells, he asks, " Pourquoi
d'ailleurs seroient-ils perdues dès que l'homme n'a pu
opérer leur destruction ?" (Phil. Zool., vol. i. p. 77.)
Of the influence of selection Lamarck has as little
notion, and he makes no use of the wonderful phænomena which are exhibited by domesticated animals, and
illustrate its powers.   The vast influence of Cuvier was
employed against the Lamarckian views, and, as the
untenability of some of his conclusions was easily
shown, his doctrines sank under the opprobium of
scientific, as well as of theological, heterodoxy.   Nor
have the efforts made of late years to revive them
tended to re-establish their credit in the minds of sound
thinkers acquainted with the facts of the case ; indeed
it may be doubted whether Lamarck has not suffered
more from his friends than from his foes.

Two years ago, in fact, though we venture to question
if even the strongest supporters of the special creation
hypothesis had not, now and then, an uneasy consciousness that all was not right, their position seemed more
impregnable than ever, if not by its own inherent strength,
at any rate by the obvious failure of all the attempts
which had been made to carry it.   On the other hand,
however much the few, who thought deeply on the
question of species, might be repelled by the generally
received dogmas, they saw no way of escaping from

them, save by the adoption of suppositions, so little justified by experiment or by observation, as to be at least equally distasteful.

The choice lay between two absurdities and a middle condition of uneasy scepticism; which last, however unpleasant and unsatisfactory, was obviously the only justifiable state of mind under the circumstances.

Such being the general ferment in the minds of naturalists, it is no wonder that they mustered strong in the rooms of the Linnæan Society, on the 1st of July of the year 1858, to hear two papers by authors living on opposite sides of the globe, working out their results independently, and yet professing to have discovered one and the same solution of all the problems connected with species. The one of these authors was an able naturalist, Mr. Wallace, who had been employed for some years in studying the productions of the islands of the Indian Archipelago, and who had forwarded a memoir embodying his views to Mr. Darwin, for communication to the Linnæan Society. On perusing the essay, Mr. Darwin was not a little surprised to find that it embodied some of the leading ideas of a great work which he had been preparing for twenty years, and parts of which, containing a development of the very same views, had been perused by his private friends fifteen or sixteen years before. Perplexed in what manner to do full justice both to his friend and to himself, Mr. Darwin placed the matter in the hands of Dr. Hooker and Sir Charles Lyell, by whose advice he communicated a brief abstract of his own views to the Linnæan Society, at the same time that Mr. Wallace's paper was read. Of that abstract, the work on the "Origin of Species" is an enlargement :

but a complete statement of Mr. Darwin's doctrine is looked for in the large and well-illustrated work which he is said to be preparing for publication.

The Darwinian hypothesis has the merit of being eminently simple and comprehensible in principle, and its essential positions may be stated in a very few words : all species have been produced by the development of varieties from common stocks by the conversion of these first into permanent races and then into new species, by the process of *natural selection*, which process is essentially identical with that artificial selection by which man has originated the races of domestic animals—the *struggle for existence* taking the place of man, and exerting, in the case of natural selection, that selective action which he performs in artificial selection.

The evidence brought forward by Mr. Darwin in support of his hypothesis is of three kinds. First, he endeavours to prove that species may be originated by selection ; secondly, he attempts to show that natural causes are competent to exert selection ; and thirdly, he tries to prove that the most remarkable and apparently anomalous phænomena exhibited by the distribution, development, and mutual relations of species, can be shown to be deducible from the general doctrine of their origin, which he propounds, combined with the known facts of geological change ; and that, even if all these phænomena are not at present explicable by it, none are necessarily inconsistent with it.

There cannot be a doubt that the method of inquiry which Mr. Darwin has adopted is not only rigorously in accordance with the canons of scientific logic, but that it

is the only adequate method.   Critics exclusively trained
in classics or in mathematics, who have never determined a
scientific fact in their lives by induction from experiment
or observation, prate learnedly about Mr. Darwin's
method, which is not inductive enough, not Baconian
enough, forsooth, for them.   But even if practical acquaintance with the process of scientific investigation is
denied them, they may learn, by the perusal of Mr.
Mill's admirable chapter " On the Deductive Method,"
that there are multitudes of scientific inquiries, in which
the method of pure induction helps the investigator but
a very little way.

" The mode of investigation," says Mr. Mill, "which, from the
proved inapplicability of direct methods of observation and experiment, remains to us as the main source of the knowledge we possess,
or can acquire, respecting the conditions and laws of recurrence of the
more complex phænomena, is called, in its most general expression,
the deductive method, and consists of three operations : the first, one
of direct induction ; the second, of ratiocination ; and the third, of
verification."

Now, the conditions which have determined the existence of species are not only exceedingly complex,
but, so far as the great majority of them are concerned,
are necessarily beyond our cognizance.   But what Mr.
Darwin has attempted to do is in exact accordance with
the rule laid down by Mr. Mill ; he has endeavoured to
determine certain great facts inductively, by observation
and experiment ; he has then reasoned from the data
thus furnished ; and lastly, he has tested the validity of
his ratiocination by comparing his deductions with the
observed facts of Nature.   Inductively, Mr. Darwin endeavours to prove that species arise in a given way.
Deductively, he desires to show that, if they arise in that

way, the facts of distribution, development, classification, &c., may be accounted for, *i. e.* may be deduced from their mode of origin, combined with admitted changes in physical geography and climate, during an indefinite period.   And this explanation, or coincidence of observed with deduced facts, is, so far as it extends, a verification of the Darwinian view.

There is no fault to be found with Mr. Darwin's method, then; but it is another question whether he has fulfilled all the conditions imposed by that method.   Is it satisfactorily proved, in fact, that species may be originated by selection? that there is such a thing as natural selection? that none of the phænomena exhibited by species are inconsistent with the origin of species in this way?   If these questions can be answered in the affirmative, Mr. Darwin's view steps out of the ranks of hypotheses into those of proved theories; but, so long as the evidence at present adduced falls short of enforcing that affirmation, so long, to our minds, must the new doctrine be content to remain among the former—an extremely valuable, and in the highest degree probable, doctrine, indeed the only extant hypothesis which is worth anything in a scientific point of view; but still a hypothesis, and not yet the theory of species.

After much consideration, and with assuredly no bias against Mr. Darwin's views, it is our clear conviction that, as the evidence stands, it is not absolutely proven that a group of animals, having all the characters exhibited by species in Nature, has ever been originated by selection, whether artificial or natural.   Groups having the morphological character of species, distinct and permanent races in fact, have been so produced over and over again ; but there is no positive evidence, at present, that any group of animals has, by variation and selective breeding, given rise to another group which was even in the least degree infertile with the first. Mr. Darwin is perfectly aware of this weak point, and brings forward a multitude of ingenious and important arguments to diminish the force of the objection. We admit the value of these arguments to their fullest extent; nay, we will go so far as to express our belief that experiments, conducted by a skilful physiologist, would very probably obtain the desired production of mutually more or less infertile breeds from a common stock, in a comparatively few years; but still, as the case stands at present, this "little rift within the lute" is not to be disguised nor overlooked.

In the remainder of Mr. Darwin's argument our own private ingenuity has not hitherto enabled us to pick holes of any great importance ; and judging by what we hear and read, other adventurers in the same field do not seem to have been much more fortunate. It has been urged, for instance, that in his chapters on the struggle for existence and on natural selection, Mr. Darwin does not so much prove that natural selection does occur, as that it must occur ; but, in fact, no other sort of demonstration is attainable. A race does not attract our attention in Nature until it has, in all probability, existed for a considerable time, and then it is too late to inquire into the conditions of its origin. Again, it is said that there is no real analogy between the selection which takes place under domestication, by human influence, and any operation which can be effected by Nature, for man interferes intelligently. Reduced to its elements, this argument implies that an effect produced with trouble by an intelligent agent must, *à fortiori*, be more troublesome, if not impossible, to an unintelligent agent. Even putting aside the question whether Nature, acting as she does according to definite and invariable laws, can be rightly called an unintelligent agent, such a position as this is wholly untenable. Mix salt and sand, and it shall puzzle the wisest of men, with his mere natural appliances, to separate all the grains of sand from all the grains of salt; but a shower of rain will effect the same object in ten minutes. And so, while man may find it tax all his intelligence to separate any variety which arises, and to breed selectively from it, the destructive agencies incessantly at work in Nature, if they find one variety to be more soluble in circumstances than the other, will inevitably, in the long run, eliminate it.

A frequent and a just objection to the Lamarckian hypothesis of the transmutation of species is based upon the absence of transitional forms between many species. But against the Darwinian hypothesis this argument has no force. Indeed, one of the most valuable and suggestive parts of Mr. Darwin's work is that in which he proves, that the frequent absence of transitions is a necessary consequence of his doctrine, and that the stock whence two or more species have sprung, need in no respect be intermediate between these species. If any two species have arisen from a common stock in the same way as the carrier and the pouter, say, have arisen from the rock-pigeon, then the common stock of these two species need be no more intermediate between the two than the rock-pigeon is between the carrier and pouter. Clearly appreciate the force of this analogy, and all the arguments against the origin of species by selection, based on the absence of transitional forms, fall to the ground. And Mr. Darwin's position might, we think, have been even stronger than it is if he had not embarrassed himself with the aphorism, "*Natura non facit saltum*," which turns up so often in his pages. We believe, as we have said above, that Nature does make jumps now and then, and a recognition of the fact is of no small importance in disposing of many minor objections to the doctrine of transmutation.

But we must pause. The discussion of Mr. Darwin's arguments in detail would lead us far beyond the limits within which we proposed, at starting, to confine this article. Our object has been attained if we have given an intelligible, however brief, account of the established facts connected with species, and of the relation of the explanation of those facts offered by Mr. Darwin to the theoretical views held by his predecessors and his contemporaries, and, above all, to the requirements of scientific logic. We have ventured to point out that it does not, as yet, satisfy all those requirements ; but we do not hesitate to assert that it is as superior to any preceding or contemporary hypothesis, in the extent of observational and experimental basis on which it rests, in its rigorously scientific method, and in its power of explaining biological phænomena, as was the hypothesis of Copernicus to the speculations of Ptolemy. But the planetary orbits turned out to be not quite circular after all, and, grand as was the service Copernicus rendered to science, Kepler and Newton had to come after him. What if the orbit of Darwinism should be a little too circular ? What if species should offer residual phænomena, here and there,

not explicable by natural selection ?   Twenty years hence
naturalists may be in a position to say whether this is, or
is not, the case; but in either event they will owe the
author of "The Origin of Species" an immense debt of
gratitude.   We should leave a very wrong impression on
the reader's mind if we permitted him to suppose that
the value of that work depends wholly on the ultimate
justification of the theoretical views which it contains.
On the contrary, if they were disproved to-morrow, the
book would still be the best of its kind—the most compendious statement of well-sifted facts bearing on the
doctrine of species that has ever appeared.   The chapters
on Variation, on the Struggle for Existence, on Instinct,
on Hybridism, on the Imperfection of the Geological
Record, on Geographical Distribution, have not only no
equals, but, so far as our knowledge goes, no competitors,
within the range of biological literature.   And viewed as
a whole, we do not believe that, since the publication of
Von Baer's Researches on Development, thirty years ago,
any work has appeared calculated to exert so large an
influence, not only on the future of Biology, but in extending the domination of Science over regions of thought
into which she has, as yet, hardly penetrated.

XIII.

CRITICISMS ON "THE ORIGIN OF SPECIES."

1. UEBER DIE DARWIN'SCHE SCHÖPFUNGSTHEORIE; EIN VORTRAG, VON A. KÖLLIKER. Leipzig, 1864.
2. EXAMINATION DU LIVRE DE M. DARWIN SUR L'ORIGINE DES ESPÈCES. Par P. FLOURENS. Paris, 1864.

IN the course of the present year [1864] several foreign commentaries upon Mr. Darwin's great work have made their appearance. Those who have perused that remarkable chapter of the " Antiquity of Man," in which Sir Charles Lyell draws a parallel between the development of species and that of languages, will be glad to hear that one of the most eminent philologers of Germany, Professor Schleicher, has, independently, published a most instructive and philosophical pamphlet (an excellent notice of which is to be found in the *Reader*, for February 27th of this year) supporting similar views with all the weight of his special knowledge and established authority as a linguist. Professor Haeckel, to whom Schleicher addresses himself, previously took occasion, in his splendid monograph on the *Radiolaria*,[1]

---

[1] " Die Radiolarien : eine Monographie," p. 231.

to express his high appreciation of, and general concord-
ance with, Mr. Darwin's views.

But the most elaborate criticisms of the " Origin of
Species" which have appeared are two works of very
widely different merit, the one by Professor Kölliker, the
well-known anatomist and histologist of Würzburg ; the
other by M. Flourens, Perpetual Secretary of the French
Academy of Sciences.

Professor Kölliker's critical essay " Upon the Darwinian Theory " is, like all that proceeds from the pen
of that thoughtful and accomplished writer, worthy of
the most careful consideration.   It comprises a brief but
clear sketch of Darwin's views, followed by an enume-
ration of the leading difficulties in the way of their
acceptance ; difficulties which would appear to be insurmountable to Professor Kölliker, inasmuch as he proposes
to replace Mr. Darwin's Theory by one which he terms
the " Theory of Heterogeneous Generation."   We shall
proceed to consider first the destructive, and secondly,
the constructive portion of the essay.

We regret to find ourselves compelled to dissent very
widely from many of Professor Kölliker's remarks ; and
from none more thoroughly than from those in which he
seeks to define what we may term the philosophical
position of Darwinism.

"Darwin," says Professor Kölliker, "is, in the fullest sense of the
word, a Teleologist.   He says quite distinctly (First Edition, pp. 199,
200) that every particular in the structure of an animal has been
created for its benefit, and he regards the whole series of animal forms
only from this point of view."

And again :

" 7. The teleological general conception adopted by Darwin is a
mistaken one.

" Varieties arise irrespectively of the notion of purpose, or of utility, according to general laws of Nature, and may be either useful, or hurtful, or indifferent.

" The assumption that an organism exists only on account of some definite end in view, and represents something more than the incorporation of a general idea, or law, implies a one-sided conception of the universe. Assuredly, every organ has, and every organism fulfils, its end, but its purpose is not the condition of its existence. Every organism is also sufficiently perfect for the purpose it serves, and in that, at least, it is useless to seek for a cause of its improvement."

It is singular how differently one and the same book will impress different minds. That which struck the present writer most forcibly on his first perusal of the " Origin of Species " was the conviction that Teleology, as commonly understood, had received its deathblow at Mr. Darwin's hands. For the teleological argument runs thus : an organ or organism (A) is precisely fitted to perform a function or purpose (B) ; therefore it was specially constructed to perform that function. In Paley's famous illustration, the adaptation of all the parts of the watch to the function, or purpose, of showing the time, is held to be evidence that the watch was specially contrived to that end ; on the ground, that the only cause we know of, competent to produce such an effect as a watch which shall keep time, is a contriving intelligence adapting the means directly to that end.

Suppose, however, that any one had been able to show that the watch had not been made directly by any person, but that it was the result of the modification of another watch which kept time but poorly ; and that this again had proceeded from a structure which could hardly be called a watch at all—seeing that it had no figures on the dial and the hands were rudimentary; and that going back and back in time we came at last to a revolving barrel as the earliest traceable rudiment of the whole fabric. And imagine that it had been possible to show that all these changes had resulted, first, from a tendency of the structure to vary indefinitely ; and secondly, from something in the surrounding world which helped all variations in the direction of an accurate time-keeper, and checked all those in other directions ; then it is obvious that the force of Paley's argument would be gone. For it would be demonstrated that an apparatus thoroughly well adapted to a particular purpose might be the result of a method of trial and error worked by unintelligent agents, as well as of the direct application of the means appropriate to that end, by an intelligent agent.

Now it appears to us that what we have here, for illustration's sake, supposed to be done with the watch, is exactly what the establishment of Darwin's Theory will do for the organic world. For the notion that every organism has been created as it is and launched straight at a purpose, Mr. Darwin substitutes the conception of something which may fairly be termed a method of trial and error. Organisms vary incessantly ; of these variations the few meet with surrounding conditions which suit them and thrive ; the many are unsuited and become extinguished.

According to Teleology, each organism is like a rifle bullet fired straight at a mark ; according to Darwin, organisms are like grapeshot of which one hits something and the rest fall wide.

For the teleologist an organism exists because it was made for the conditions in which it is found ; for the Darwinian an organism exists because, out of many of

its kind, it is the only one which has been able to persist
in the conditions in which it is found.

Teleology implies that the organs of every organism
are perfect and cannot be improved; the Darwinian
theory simply affirms that they work well enough to
enable the organism to hold its own against such competitors as it has met with, but admits the possibility of
indefinite improvement. But an example may bring
into clearer light the profound opposition between the
ordinary teleological, and the Darwinian, conception.

Cats catch mice, small birds and the like, very well.
Teleology tells us that they do so because they were
expressly constructed for so doing—that they are perfect
mousing apparatuses, so perfect and so delicately ad-
justed that no one of their organs could be altered,
without the change involving the alteration of all the
rest. Darwinism affirms, on the contrary, that there
was no express construction concerned in the matter;
but that among the multitudinous variations of the
Feline stock, many of which died out from want of
power to resist opposing influences, some, the cats, were
better fitted to catch mice than others, whence they
throve and persisted, in proportion to the advantage
over their fellows thus offered to them.

Far from imagining that cats exist *in order* to catch
mice well, Darwinism supposes that cats exist *because*
they catch mice well—mousing being not the end, but
the condition, of their existence. And if the cat-type
has long persisted as we know it, the interpretation of
the fact upon Darwinian principles would be, not that
the cats have remained invariable, but that such varieties
as have incessantly occurred have been, on the whole,

less fitted to get on in the world than the existing stock.

If we apprehend the spirit of the " Origin of Species " rightly, then, nothing can be more entirely and absolutely opposed to Teleology, as it is commonly understood, than the Darwinian Theory. So far from being a " Teleologist in the fullest sense of the word," we should deny that he is a Teleologist in the ordinary sense at all ; and we should say that, apart from his merits as a naturalist, he has rendered a most remarkable service to philosophical thought by enabling the student of Nature to recognise, to their fullest extent, those adaptations to purpose which are so striking in the organic world, and which Teleology has done good service in keeping before our minds, without being false to the fundamental principles of a scientific conception of the universe. The apparently diverging teachings of the Teleologist and of the Morphologist are reconciled by the Darwinian hypothesis.

But leaving our own impressions of the " Origin of Species," and turning to those passages specially cited by Professor Kölliker, we cannot admit that they bear the interpretation he puts upon them. Darwin, if we read him rightly, does *not* affirm that every detail in the structure of an animal has been created for its benefit. His words are (p. 199) :—

"The foregoing remarks lead me to say a few words on the protest lately made by some naturalists against the utilitarian doctrine that every detail of structure has been produced for the good of its possessor. They believe that very many structures have been created for beauty in the eyes of man, or for mere variety. This doctrine, if true, would be absolutely fatal to my theory—yet I fully admit that many structures are of no direct use to their possessor."

And after sundry illustrations and qualifications, he concludes (p. 200) :—

" Hence every detail of structure in every living creature (making some little allowance for the direct action of physical conditions) may be viewed either as having been of special use to some ancestral form, or as being now of special use to the descendants of this form—either directly, or indirectly, through the complex laws of growth."

But it is one thing to say, Darwinically, that every detail observed in an animal's structure is of use to it, or has been of use to its ancestors ; and quite another to affirm, teleologically, that every detail of an animal's structure has been created for its benefit. On the former hypothesis, for example, the teeth of the fœtal *Balæna* have a meaning ; on the latter, none. So far as we are aware, there is not a phrase in the " Origin of Species," inconsistent with Professor Kölliker's position, that " varieties arise irrespectively of the notion of purpose, or of utility, according to general laws of Nature, and may be either useful, or hurtful, or indifferent."

On the contrary, Mr. Darwin writes (Summary of Chap. V.) :—

" Our ignorance of the laws of variation is profound. Not in one case out of a hundred can we pretend to assign any reason why this or that part varies more or less from the same part in the parents. . . . . The external conditions of life, as climate and food, &c. seem to have induced some slight modifications. Habit, in producing constitutional differences, and use, in strengthening, and disuse, in weakening and diminishing organs, seem to have been more potent in their effects."

And finally, as if to prevent all possible misconception, Mr. Darwin concludes his Chapter on Variation with these pregnant words :—

" Whatever the cause may be of each slight difference in the offspring from their parents—and a cause for each must exist—it is the steady

accumulation, through natural selection of such differences, when beneficial to the individual, that gives rise to all the more important
modifications of structure, by which the innumerable beings on the
face of the earth are enabled to struggle with each other, and the best
adapted to survive."

We have dwelt at length upon this subject, because of
its great general importance, and because we believe that
Professor Kölliker's criticisms on this head are based
upon a misapprehension of Mr. Darwin's views—substantially they appear to us to coincide with his own.
The other objections which Professor Kölliker enumerates
and discusses are the following :[1]—

" 1. No transitional forms between existing species are known ;
and known varieties, whether selected or spontaneous, never go so
far as to establish new species."

To this Professor Kölliker appears to attach some
weight. He makes the suggestion that the short-faced
tumbler pigeon may be a pathological product.

" 2. No transitional forms of animals are met with among the
organic remains of earlier epochs."

Upon this, Professor Kölliker remarks that the absence
of transitional forms in the fossil world, though not necessarily fatal to Darwin's views, weakens his case.

" 3. The struggle for existence does not take place."

To this objection, urged by Pelzeln, Kölliker, very
justly, attaches no weight.

" 4. A tendency of organisms to give rise to useful varieties, and
a natural selection, do not exist.

" The varieties which are found arise in consequence of manifold
external influences, and it is not obvious why they all, or partially,

---

[1] Space will not allow us to give Professor Kölliker's arguments in detail ;
our readers will find a full and accurate version of them in the *Reader* for
August 13th and 20th, 1864.

should be particularly useful. Each animal suffices for its own ends, is perfect of its kind, and needs no further development. Should, however, a variety be useful and even maintain itself, there is no obvious reason why it should change any further. The whole conception of the imperfection of organisms and the necessity of their becoming perfected is plainly the weakest side of Darwin's Theory, and a *pis aller* (Nothbehelf) because Darwin could think of no other principle by which to explain the metamorphoses which, as I also believe, have occurred."

Here again we must venture to dissent completely from Professor Kölliker's conception of Mr. Darwin's hypothesis. It appears to us to be one of the many peculiar merits of that hypothesis that it involves no belief in a necessary and continual progress of organisms.

Again, Mr. Darwin, if we read him aright, assumes no special tendency of organisms to give rise to useful varieties, and knows nothing of needs of development, or necessity of perfection. What he says is, in substance : All organisms vary. It is in the highest degree improbable that any given variety should have exactly the same relations to surrounding conditions as the parent stock. In that case it is either better fitted (when the variation may be called useful), or worse fitted, to cope with them. If better, it will tend to supplant the parent stock ; if worse, it will tend to be extinguished by the parent stock.

If (as is hardly conceivable) the new variety is so perfectly adapted to the conditions that no improvement upon it is possible,—it will persist, because, though it does not cease to vary, the varieties will be inferior to itself.

If, as is more probable, the new variety is by no means perfectly adapted to its conditions, but only fairly well adapted to them, it will persist, so long as none of the varieties which it throws off are better adapted than itself.

On the other hand, as soon as it varies in a useful way, *i. e.* when the variation is such as to adapt it more perfectly to its conditions, the fresh variety will tend to supplant the former.

So far from a gradual progress towards perfection forming any necessary part of the Darwinian creed, it appears to us that it is perfectly consistent with indefinite persistence in one state, or with a gradual retrogression. Suppose, for example, a return of the glacial epoch and a spread of polar climatal conditions over the whole globe. The operation of natural selection under these circumstances would tend, on the whole, to the weeding out of the higher organisms and the cherishing of the lower forms of life. Cryptogamic vegetation would have the advantage over Phanerogamic; *Hydrozoa* over Corals; *Crustacea* over *Insecta,* and *Amphipoda* and *Isopoda* over the higher *Crustacea;* Cetaceans and Seals over the *Primates;* the civilization of the Esquimaux over that of the European.

" 5. Pelzeln has also objected that if the later organisms have proceeded from the earlier, the whole developmental series, from the simplest to the highest, could not now exist; in such a case the simpler organisms must have disappeared."

To this Professor Kölliker replies, with perfect justice, that the conclusion drawn by Pelzeln does not really follow from Darwin's premises, and that, if we take the facts of Palæontology as they stand, they rather support than oppose Darwin's theory.

" 6. Great weight must be attached to the objection brought forward by Huxley, otherwise a warm supporter of Darwin's hypothesis, that we know of no varieties which are sterile with one another, as is the rule among sharply distinguished animal forms.

"If Darwin is right, it must be demonstrated that forms may be produced by selection, which, like the present sharply distinguished animal forms, are infertile when coupled with one another, and this has not been done."

The weight of this objection is obvious; but our ignorance of the conditions of fertility and sterility, the want of carefully conducted experiments extending over long series of years, and the strange anomalies presented by the results of the cross-fertilization of many plants, should all, as Mr. Darwin has urged, be taken into account in considering it.

The seventh objection is that we have already discussed (*suprà*, p. 329).

The eighth and last stands as follows:—

"8. The developmental theory of Darwin is not needed to enable us to understand the regular harmonious progress of the complete series of organic forms from the simpler to the more perfect.

"The existence of general laws of Nature explains this harmony, even if we assume that all beings have arisen separately and independent of one another. Darwin forgets that inorganic nature, in which there can be no thought of a genetic connexion of forms, exhibits the same regular plan, the same harmony, as the organic world; and that, to cite only one example, there is as much a natural system of minerals as of plants and animals."

We do not feel quite sure that we seize Professor Kölliker's meaning here, but he appears to suggest that the observation of the general order and harmony which pervade inorganic nature, would lead us to anticipate a similar order and harmony in the organic world. And this is no doubt true, but it by no means follows that the particular order and harmony observed among them should be that which we see. Surely the stripes of dun

horses, and the teeth of the fœtal *Balæna*, are not explained by the " existence of general laws of Nature."
Mr. Darwin endeavours to explain the exact order of
organic nature which exists ; not the mere fact that
there is some order.

And with regard to the existence of a natural system
of minerals ; the obvious reply is that there may be a
natural classification of any objects—of stones on a sea-
beach, or of works of art ; a natural classification being
simply an assemblage of objects in groups, so as to
express their most important and fundamental resemblances and differences.   No doubt Mr. Darwin believes that those resemblances and differences upon
which our natural systems or classifications of animals
and plants are based, are resemblances and differences
which have been produced genetically, but we can discover no reason for supposing that he denies the existence
of natural classifications of other kinds.

And, after all, is it quite so certain that a genetic
relation may not underlie the classification of minerals ?
The inorganic world has not always been what we see
it.   It has certainly had its metamorphoses, and, very
probably, a long " Entwickelungsgeschichte" out of a
nebular blastema.   Who knows how far that amount of
likeness among sets of minerals, in virtue of which they
are now grouped into families and orders, may not be
the expression of the common conditions to which that
particular patch of nebulous fog, which may have been
constituted by their atoms, and of which they may be,
in the strictest sense, the descendants, was subjected ?

It will be obvious from what has preceded, that we
do not agree with Professor Kölliker in thinking the

objections which he brings forward so weighty as to be
fatal to Darwin's view. But even if the case were otherwise, we should be unable to accept the " Theory of
Heterogeneous Generation" which is offered as a substitute. That theory is thus stated :—

> " The fundamental conception of this hypothesis is, that, under the
> influence of a general law of development, the germs of organisms
> produce others different from themselves. This might happen (1) by
> the fecundated ova passing, in the course of their development, under
> particular circumstances, into higher forms ; (2) by the primitive and
> later organisms producing other organisms without fecundation, out of
> germs or eggs (Parthenogenesis)."

In favour of this hypothesis, Professor Kölliker adduces the well-known facts of Agamogenesis, or " alternate generation ; " the extreme dissimilarity of the
males and females of many animals ; and of the males,
females, and neuters of those insects which live in
colonies : and he defines its relations to the Darwinian
theory as follows :—

> " It is obvious that my hypothesis is apparently very similar to
> Darwin's, inasmuch as I also consider that the various forms of
> animals have proceeded directly from one another. My hypothesis of
> the creation of organisms by heterogeneous generation, however, is
> distinguished very essentially from Darwin's by the entire absence of
> the principle of useful variations and their natural selection ; and my
> fundamental conception is this, that a great plan of development lies
> at the foundation of the origin of the whole organic world, impelling
> the simpler forms to more and more complex developments. How
> this law operates, what influences determine the development of the
> eggs and germs, and impel them to assume constantly new forms, I
> naturally cannot pretend to say ; but I can at least adduce the great
> analogy of the alternation of generations. If a *Bipinnaria,* a *Brachi-*
> *alaria,* a *Pluteus,* is competent to produce the Echinoderm, which is
> so widely different from it ; if a hydroid polype can produce the higher
> Medusa ; if the vermiform Trematode ' nurse ' can develop within
> itself the very unlike *Cercaria,* it will not appear impossible that the

egg, or ciliated embryo, of a sponge, for once, under special conditions, might become a hydroid polype, or the embryo of a Medusa, an Echinoderm."

It is obvious, from these extracts, that Professor Kölliker's hypothesis is based upon the supposed existence of a close analogy between the phænomena of Agamogenesis and the production of new species from pre-existing ones. But is the analogy a real one? We think that it is not, and, by the hypothesis, cannot be.

For what are the phænomena of Agamogenesis, stated generally? An impregnated egg develops into an asexual form, A; this gives rise, asexually, to a second form or forms, B, more or less different from A. B may multiply asexually again; in the simpler cases, however, it does not, but, acquiring sexual characters, produces impregnated eggs from whence A once more arises.

No case of Agamogenesis is known in which, *when A differs widely from B*, it is itself capable of sexual propagation. No case whatever is known in which the progeny of B, by sexual generation, is other than a reproduction of A.

But if this be a true statement of the nature of the process of Agamogenesis, how can it enable us to comprehend the production of new species from already existing ones? Let us suppose Hyænas to have preceded Dogs, and to have produced the latter in this way. Then the Hyæna will represent A, and the Dog, B. The first difficulty that presents itself is that the Hyæna must be asexual, or the process will be wholly without analogy in the world of Agamogenesis. But passing over this difficulty, and supposing a male and female Dog to be produced at the same time from the

Hyæna stock, the progeny of the pair, if the analogy of the simpler kinds of Agamogenesis[1] is to be followed, should be a litter, not of puppies, but of young Hyænas. For the Agamogenetic series is always, as we have seen, A : B : A : B, &c.; whereas, for the production of a new species, the series must be A : B : B : B, &c. The production of new species, or genera, is the extreme permanent divergence from the primitive stock. All known Agamogenetic processes, on the other hand, end in a complete return to the primitive stock. How then is the production of new species to be rendered intelligible by the analogy of Agamogenesis ?

The other alternative put by Professor Kölliker—the passage of fecundated ova in the course of their development into higher forms—would, if it occurred, be merely an extreme case of variation in the Darwinian sense, greater in degree than, but perfectly similar in kind to, that which occurred when the well-known Ancon Ram was developed from an ordinary Ewe's ovum. Indeed we have always thought that Mr. Darwin has unnecessarily hampered himself by adhering so strictly to his favourite "Natura non facit saltum." We greatly suspect that she does make considerable jumps in the way of variation now and then, and that these

---

[1] If, on the contrary, we follow the analogy of the more complex forms of Agamogenesis, such as that exhibited by some *Trematoda* and by the *Aphides*, the Hyæna must produce, asexually, a brood of asexual Dogs, from which other sexless Dogs must proceed. At the end of a certain number of terms of the series, the Dogs would acquire sexes and generate young ; but these young would be, not Dogs, but Hyænas. In fact, we have *demonstrated*, in Agamogenetic phænomena, that inevitable recurrence to the original type, which is *asserted* to be true of variations in general, by Mr. Darwin's opponents ; and which, if the assertion could be changed into a demonstration, would, in fact, be fatal to his hypothesis.

saltations give rise to some of the gaps which appear to exist in the series of known forms.

Strongly and freely as we have ventured to disagree with Professor Kölliker, we have always done so with regret, and we trust without violating that respect which is due, not only to his scientific eminence and to the careful study which he has devoted to the subject, but to the perfect fairness of his argumentation, and the generous appreciation of the worth of Mr. Darwin's labours which he always displays. It would be satisfactory to be able to say as much for M. Flourens.

But the Perpetual Secretary of the French Academy of Sciences deals with Mr. Darwin as the first Napoleon would have treated an "idéologue ;" and while displaying a painful weakness of logic and shallowness of information, assumes a tone of authority, which always touches upon the ludicrous, and sometimes passes the limits of good breeding.

For example (p. 56) :—

> "M. Darwin continue : 'Aucune distinction absolue n'a été et ne peut être établie entre les espèces et les variétés.' Je vous ai déjà dit que vous vous trompiez ; une distinction absolue sépare les variétés d'avec les espèces."

"*Je vous ai déjà dit ;* moi, M. le Secrétaire perpétuel de l'Académie des Sciences : et vous

> 'Qui n'êtes rien,
> Pas même Académicien ;'

what do you mean by asserting the contrary ?" Being devoid of the blessings of an Academy in England, we are unaccustomed to see our ablest men treated in this fashion even by a " Perpetual Secretary."

Or again, considering that if there is any one quality of Mr. Darwin's work to which friends and foes have alike borne witness, it is his candour and fairness in admitting and discussing objections, what is to be thought of M. Flourens' assertion, that

"M. Darwin ne cite que les auteurs qui partagent ses opinions." (P. 40.)

Once more (p. 65) :

"Enfin l'ouvrage de M. Darwin a paru. On ne peut qu'être frappé du talent de l'auteur. Mais que d'idées obscures, que d'idées fausses! Quel jargon métaphysique jeté mal à propos dans l'histoire naturelle, qui tombe dans le galimatias dès qu'elle sort des idées claires, des idées justes! Quel langage prétentieux et vide! Quelles personifications puériles et surannées! O lucidité! O solidité de l'esprit Français, que devenez-vous ?"

"Obscure ideas," " metaphysical jargon," " pretentious and empty language," " puerile and superannuated personifications."   Mr. Darwin has many and hot opponents on this side of the Channel and in Germany, but we do not recollect to have found precisely these sins in the long catalogue of those hitherto laid to his charge.   It is worth while, therefore, to examine into these discoveries effected solely by the aid of the " lucidity and solidity " of the mind of M. Flourens.

According to M. Flourens, Mr. Darwin's great error is that he has personified Nature (p. 10), and further that he has

" imagined a natural selection : he imagines afterwards that this power of selecting (*pouvoir d'élire*) which he gives to Nature is similar to the power of man. These two suppositions admitted, nothing stops him : he plays with Nature as he likes, and makes her do all he pleases." (P. 6.)

And this is the way M. Flourens extinguishes natural selection :

" Voyons donc encore une fois, ce qu'il peut y avoir de fondé dans ce qu'on nomme *élection naturelle.*

" *L'élection naturelle* n'est sous un autre nom que la nature. Pour un être organisé, la nature n'est que l'organisation, ni plus ni moins.

" Il faudra donc aussi personnifier *l'organisation,* et dire que *l'organisation* choisit *l'organisation.* *L'election naturelle* est cette *forme substantielle* dont on jouait autrefois avec tant de facilité. Aristote disait que ' Si l'art de bâtir était dans le bois, cet art agirait comme la nature.' A la place de *l'art de bâtir* M. Darwin met *l'election naturelle,* et c'est tout un : l'un n'est pas plus chimérique que l'autre." (P. 31.)

And this is really all that M. Flourens can make of Natural Selection. We have given the original, in fear lest a translation should be regarded as a travesty ; but with the original before the reader, we may try to analyse the passage. " For an organized being, Nature is only organization, neither more nor less."

Organized beings then have absolutely no relation to inorganic nature : a plant does not depend on soil or sunshine, climate, depth in the ocean, height above it ; the quantity of saline matters in water have no influence upon animal life ; the substitution of carbonic acid for oxygen in our atmosphere would hurt nobody ! That these are absurdities no one should know better than M. Flourens ; but they are logical deductions from the assertion just quoted, and from the further statement that natural selection means only that " organization chooses and selects organization."

For if it be once admitted (what no sane man denies) that the chances of life of any given organism are increased by certain conditions (A) and diminished by their opposites (B), then it is mathematically certain that any change of conditions in the direction of (A) will exercise a selective influence in favour of that organism, tending to its increase and multiplication, while any change in the direction of (B) will exercise a selective influence against that organism, tending to its decrease and extinction.

Or, on the other hand, conditions remaining the same, let a given organism vary (and no one doubts that they do vary) in two directions : into one form (*a*) better fitted to cope with these conditions than the original stock, and a second (*b*) less well adapted to them.   Then it is no less certain that the conditions in question must exercise a selective influence in favour of (*a*) and against (*b*), so that (*a*) will tend to predominance, and (*b*) to extirpation.

That M. Flourens should be unable to perceive the logical necessity of these simple arguments, which lie at the foundation of all Mr. Darwin's reasoning ; that he should confound an irrefragable deduction from the observed relations of organisms to the conditions which lie around them, with a metaphysical "forme substantielle," or a chimerical personification of the powers of Nature, would be incredible, were it not that other passages of his work leave no room for doubt upon the subject.

" On imagine une *élection naturelle* que, pour plus de ménagement, on me dit être *inconsciente*, sans s'apercevoir que le contre-sens littéral est précisément là : *élection inconsciente.*"   (P. 52.)

" J'ai déjà dit ce qu'il faut penser de *l'élection naturelle.*   Ou *l'élection naturelle* n'est rien, ou c'est la nature : mais la nature douée *d'élection*, mais la nature personnifiée : dernière erreur du dernier siècle : Le xixe ne fait plus de personnifications."   (P. 53.)

M. Flourens cannot imagine an unconscious selection
—it is for him a contradiction in terms.   Did M.
Flourens ever visit one of the prettiest watering-places
of " la belle France," the Baie d'Arcachon ?   If so, he
will probably have passed through the district of the
Landes, and will have had an opportunity of observing
the formation of " dunes " on a grand scale.   What are
these " dunes ?"   The winds and waves of the Bay of
Biscay have not much consciousness, and yet they have
with great care " selected," from among an infinity of
masses of silex of all shapes and sizes, which have been
submitted to their action, all the grains of sand below a
certain size, and have heaped them by themselves over
a great area.   This sand has been " unconsciously
selected " from amidst the gravel in which it first lay
with as much precision as if man had " consciously
selected " it by the aid of a sieve.   Physical Geology is
full of such selections—of the picking out of the soft
from the hard, of the soluble from the insoluble, of the
fusible from the infusible, by natural agencies to which
we are certainly not in the habit of ascribing consciousness.

But that which wind and sea are to a sandy beach,
the sum of influences, which we term the " conditions
of existence," is to living organisms.   The weak are
sifted out from the strong.   A frosty night " selects "
the hardy plants in a plantation from among the tender
ones as effectually as if it were the wind, and they, the
sand and pebbles, of our illustration ; or, on the other
hand, as if the intelligence of a gardener had been
operative in cutting the weaker organisms down.   The
thistle, which has spread over the Pampas, to the

destruction of native plants, has been more effectually "selected" by the unconscious operation of natural conditions than if a thousand agriculturists had spent their time in sowing it.

It is one of Mr. Darwin's many great services to Biological science that he has demonstrated the significance of these facts. He has shown that—given variation and given change of conditions—the inevitable result is the exercise of such an influence upon organisms that one is helped and another is impeded; one tends to predominate, another to disappear; and thus the living world bears within itself, and is surrounded by, impulses towards incessant change.

But the truths just stated are as certain as any other physical laws, quite independently of the truth, or falsehood, of the hypothesis which Mr. Darwin has based upon them; and that M. Flourens, missing the substance and grasping at a shadow, should be blind to the admirable exposition of them, which Mr. Darwin has given, and see nothing there but a "dernière erreur du dernier siècle"—a personification of Nature—leads us indeed to cry with him: "O lucidité! O solidité de l'esprit Français, que devenez-vous?"

M. Flourens has, in fact, utterly failed to comprehend the first principles of the doctrine which he assails so rudely. His objections to details are of the old sort, so battered and hackneyed on this side of the Channel, that not even a Quarterly Reviewer could be induced to pick them up for the purpose of pelting Mr. Darwin over again. We have Cuvier and the mummies; M. Roulin and the domesticated animals of America; the difficulties presented by hybridism and by Palæontology;

Darwinism a *rifacciamento* of De Maillet and Lamarck ; Darwinism a system without a commencement, and its author bound to believe in M. Pouchet, &c. &c. How one knows it all by heart, and with what relief one reads at p. 65—

> " Je laisse M. Darwin ! "

But we cannot leave M. Flourens without calling our readers' attention to his wonderful tenth chapter, " De la Préexistence des Germes et de l'Epigénèse," which opens thus :—

> " Spontaneous generation is only a chimæra. This point established, two hypotheses remain : that of *pre-existence* and that of *epigenesis.* The one of these hypotheses has as little foundation as the other." (P. 163.)

> " The doctrine of *epigenesis* is derived from Harvey : following by ocular inspection the development of the new being in the Windsor does, he saw each part appear successively, and taking the moment of *appearance* for the moment of *formation* he imagined *epigenesis.*" (P. 165.)

On the contrary, says M. Flourens (p. 167),

> " The new being is formed at a stroke (*tout d'un coup*), as a whole, instantaneously ; it is not formed part by part, and at different times. It is formed at once ; it is formed at the single *individual* moment at which the conjunction of the male and female elements takes place."

It will be observed that M. Flourens uses language which cannot be mistaken. For him, the labours of Von Baer, of Rathke, of Coste, and their contemporaries and successors in Germany, France, and England, are non-existent ; and, as Darwin *"imagina"* natural selection, so Harvey *"imagina"* that doctrine which gives him an even greater claim to the veneration of posterity than his better known discovery of the circulation of the blood.

Language such as that we have quoted is, in fact, so preposterous, so utterly incompatible with anything but absolute ignorance of some of the best established facts, that we should have passed it over in silence had it not appeared to afford some clue to M. Flourens' unhesitating, *à priori*, repudiation of all forms of the doctrine of the progressive modification of living beings. He whose mind remains uninfluenced by an acquaintance with the phænomena of development, must indeed lack one of the chief motives towards the endeavour to trace a genetic relation between the different existing forms of life. Those who are ignorant of Geology, find no difficulty in believing that the world was made as it is ; and the shepherd, untutored in history, sees no reason to regard the green mounds which indicate the site of a Roman camp, as aught but part and parcel of the primæval hill-side. So M. Flourens, who believes that embryos are formed " tout d'un coup," naturally finds no difficulty in conceiving that species came into existence in the same way.

XIV.

ON DESCARTES' "DISCOURSE TOUCHING THE METHOD OF USING ONE'S REASON RIGHTLY AND OF SEEKING SCIENTIFIC TRUTH."

IT has been well said that "all the thoughts of men, from the beginning of the world until now, are linked together into one great chain;" but the conception of the intellectual filiation of mankind which is expressed in these words may, perhaps, be more fitly shadowed forth by a different metaphor. The thoughts of men seem rather to be comparable to the leaves, flowers, and fruit upon the innumerable branches of a few great stems, fed by commingled and hidden roots. These stems bear the names of the half-a-dozen men, endowed with intellects of heroic force and clearness, to whom we are led, at whatever point of the world of thought the attempt to trace its history commences; just as certainly as the following up the small twigs of a tree to the branchlets which bear them, and tracing the branchlets to their supporting branches, brings us, sooner or later, to the bole.

It seems to me that the thinker who, more than any
other, stands in the relation of such a stem towards the
philosophy and the science of the modern world is René
Descartes.   I mean, that if you lay hold of any characteristic product of modern ways of thinking, either in
the region of philosophy, or in that of science, you find
the spirit of that thought, if not its form, to have been
present in the mind of the great Frenchman.

There are some men who are counted great because
they represent the actuality of their own age, and mirror
it as it is.   Such an one was Voltaire, of whom it was
epigrammatically said, "he expressed everybody's thoughts
better than anybody."[1]   But there are other men who
attain greatness because they embody the potentiality of
their own day, and magically reflect the future.   They
express the thoughts which will be everybody's two
or three centuries after them.   Such an one was
Descartes.

Born, in 1596, nearly three hundred years ago, of a
noble family in Touraine, René Descartes grew up into a
sickly and diminutive child, whose keen wit soon gained
him that title of "the Philosopher," which, in the mouths
of his noble kinsmen, was more than half a reproach.
The best schoolmasters of the day, the Jesuits, educated
him as well as a French boy of the seventeenth century
could be educated.   And they must have done their
work honestly and well, for, before his schoolboy days
were over, he had discovered that the most of what he
had learned, except in mathematics, was devoid of solid
and real value.

---

[1] I forget who it was said of him : "Il a plus que personne l'esprit que tout
le monde a."

"Therefore," says he, in that "Discourse "[1] which I have taken for my text, " as soon as I was old enough to be set free from the government of my teachers, I entirely forsook the study of letters ; and determining to seek no other knowledge than that which I could discover within myself, or in the great book of the world, I spent the remainder of my youth in travelling ; in seeing courts and armies ; in the society of people of different humours and conditions; in gathering varied experience ; in testing myself by the chances of fortune ; and in always trying to profit by my reflections on what happened. . . . And I always had an intense desire to learn how to distinguish truth from falsehood, in order to be clear about my actions, and to walk surefootedly in this life."

But "learn what is true, in order to do what is right," is the summing up of the whole duty of man, for all who are unable to satisfy their mental hunger with the east wind of authority; and to those of us moderns who are in this position, it is one of Descartes' great claims to our reverence as a spiritual ancestor, that, at three-and-twenty, he saw clearly that this was his duty, and acted up to his conviction.   At two-and-thirty, in fact, finding all other occupations incompatible with the search after the knowledge which leads to action, and being possessed of a modest competence, he withdrew into Holland ; where he spent nine years in learning and thinking, in such retirement that only one or two trusted friends knew of his whereabouts.

In 1637 the firstfruits of these long meditations were given to the world in the famous " Discourse touching the Method of using Reason rightly and of seeking scientific Truth," which, at once an autobiography and a philosophy, clothes the deepest thought in language of exquisite harmony, simplicity, and clearness.

[1] "Discours de la Méthode pour bien conduire sa Raison et chercher la Vérité dans les Sciences."

The central propositions of the whole "Discourse" are these.   There is a path that leads to truth so surely, that any one who will follow it must needs reach the goal, whether his capacity be great or small.   And there is one guiding rule by which a man may always find this path, and keep himself from straying when he has found it. This golden rule is—give unqualified assent to no propositions but those the truth of which is so clear and distinct that they cannot be doubted.

The enunciation of this great first commandment of science consecrated Doubt.   It removed Doubt from the seat of penance among the grievous sins to which it had long been condemned, and enthroned it in that high place among the primary duties, which is assigned to it by the scientific conscience of these latter days.   Descartes was the first among the moderns to. obey this commandment deliberately ; and, as a matter of religious duty, to strip off all his beliefs and reduce himself to a state of intellectual nakedness, until such time as he could satisfy himself which were fit to be worn.   He thought a bare skin healthier than the most respectable and well-cut clothing of what might, possibly, be mere shoddy.

When I say that Descartes consecrated doubt, you must remember that it was that sort of doubt which Goethe has called "the active scepticism, whose whole aim is to conquer itself;"[1] and not that other sort which is born of flippancy and ignorance, and whose aim is only to perpetuate itself, as an excuse for idleness and indifference.   But it is impossible to define what is meant by

[1] "Eine thätige Skepsis ist die, welche unablässig bemüht ist sich selbst zu überwinden, und durch geregelte Erfahrung zu einer Art von bedingter Zuverlässigkeit zu gelangen."—*Maximen und Reflexionen,* 7 Abtheilung.

scientific doubt better than in Descartes' own words. After describing the gradual progress of his negative criticism, he tells us :—

> "For all that, I did not imitate the sceptics, who doubt only for doubting's sake, and pretend to be always undecided; on the contrary, my whole intention was to arrive at certainty, and to dig away the drift and the sand until I reached the rock or the clay beneath."

And further, since no man of common sense, when he pulls down his house for the purpose of rebuilding it, fails to provide himself with some shelter while the work is in progress; so, before demolishing the spacious, if not commodious, mansion of his old beliefs, Descartes thought it wise to equip himself with what he calls "*une morale par provision,*" by which he resolved to govern his practical life until such time as he should be better instructed. The laws of this "provisional self-govern-ment" are embodied in four maxims, of which one binds our philosopher to submit himself to the laws and religion in which he was brought up; another, to act, on all those occasions which call for action, promptly and according to the best of his judgment, and to abide, without repining, by the result: a third rule is to seek happiness in limiting his desires, rather than in attempting to satisfy them; while the last is to make the search after truth the business of his life.

Thus prepared to go on living while he doubted, Descartes proceeded to face his doubts like a man. One thing was clear to him, he would not lie to himself— would, under no penalties, say, "I am sure" of that of which he was not sure; but would go on digging and delving until he came to the solid adamant; or, at worst, made sure there was no adamant. As the record of his

progress tells us, he was obliged to confess that life is full
of delusions; that authority may err; that testimony
may be false or mistaken; that reason lands us in endless fallacies; that memory is often as little trustworthy
as hope; that the evidence of the very senses may be
misunderstood; that dreams are real as long as they last,
and that what we call reality may be a long and restless
dream. Nay, it is conceivable that some powerful and
malicious being may find his pleasure in deluding us, and
in making us believe the thing which is not, every moment
of our lives. What, then, is certain? What even, if
such a being exists, is beyond the reach of his powers of
delusion? Why, the fact that the thought, the present
consciousness, exists. Our thoughts may be delusive,
but they cannot be fictitious. As thoughts, they are
real and existent, and the cleverest deceiver cannot
make them otherwise.

Thus, thought is existence. More than that, so far as
we are concerned, existence is thought, all our conceptions of existence being some kind or other of thought.
Do not for a moment suppose that these are mere
paradoxes or subtleties. A little reflection upon the
commonest facts proves them to be irrefragable truths.
For example, I take up a marble, and I find it to be
a red, round, hard, single body. We call the redness,
the roundness, the hardness, and the singleness, " qualities " of the marble; and it sounds, at first, the height of
absurdity to say that all these qualities are modes of our
own consciousness, which cannot even be conceived to
exist in the marble. But consider the redness, to begin
with. How does the sensation of redness arise? The
waves of a certain very attenuated matter, the particles

of which are vibrating with vast rapidity, but with very
different velocities, strike upon the marble, and those
which vibrate with one particular velocity are thrown off
from its surface in all directions. The optical apparatus
of the eye gathers some of these together, and gives them
such a course that they impinge upon the surface of the
retina, which is a singularly delicate apparatus, connected
with the termination of the fibres of the optic nerve.
The impulses of the attenuated matter, or ether, affect
this apparatus and the fibres of the optic nerve in a
certain way ; and the change in the fibres of the optic
nerve produces yet other changes in the brain ; and
these, in some fashion unknown to us, give rise to the
feeling, or consciousness, of redness. If the marble
could remain unchanged, and either the rate of vibration
of the ether, or the nature of the retina, could be altered,
the marble would seem not red, but some other colour.
There are many people who are what are called colour-
blind, being unable to distinguish one colour from
another. Such an one might declare our marble to be
green ; and he would be quite as right in saying that it
is green, as we are in declaring it to be red. But then,
as the marble cannot, in itself, be both green and red, at
the same time, this shows that the quality "redness"
must be in our consciousness and not in the marble.

In like manner, it is easy to see that the roundness and
the hardness are forms of our consciousness, belonging
to the groups which we call sensations of sight and
touch. If the surface of the cornea were cylindrical, we
should have a very different notion of a round body
from that which we possess now ; and if the strength of
the fabric, and the force of the muscles, of the body were

increased a hundredfold, our marble would seem to be as soft as a pellet of bread crumbs.

Not only is it obvious that all these qualities are in us, but, if you will make the attempt, you will find it quite impossible to conceive of "blueness," "roundness," and "hardness" as existing without reference to some such consciousness as our own. It may seem strange to say that even the "singleness" of the marble is relative to us ; but extremely simple experiments will show that such is veritably the case, and that our two most trustworthy senses may be made to contradict one another on this very point. Hold the marble between the finger and thumb, and look at it in the ordinary way. Sight and touch agree that it is single. Now squint, and sight tells you that there are two marbles, while touch asserts that there is only one. Next, return the eyes to their natural position, and, having crossed the forefinger and the middle finger, put the marble between their tips. Then touch will declare that there are two marbles, while sight says that there is only one ; and touch claims our belief, when we attend to it, just as imperatively as sight does.

But it may be said, the marble takes up a certain space which could not be occupied, at the same time, by anything else. In other words, the marble has the primary quality of matter, extension. Surely this quality must be in the thing, and not in our minds ? But the reply must still be ; whatever may, or may not, exist in the thing, all that we can know of these qualities is a state of consciousness. What we call extension is a consciousness of a relation between two, or more, affections of the sense of sight, or of touch. And it is wholly incon-

ceivable that what we call extension should exist independently of such consciousness as our own. Whether,
notwithstanding this inconceivability, it does so exist, or
not, is a point on which I offer no opinion.

Thus, whatever our marble may be in itself, all that
we can know of it is under the shape of a bundle of our
own consciousnesses.

Nor is our knowledge of anything we know or feel
more, or less, than a knowledge of states of consciousness.
And our whole life is made up of such states. Some of
these states we refer to a cause we call " self ; " others to
a cause or causes which may be comprehended under
the title of "not-self." But neither of the existence of
" self," nor of that of "not-self," have we, or can we by
any possibility have, any such unquestionable and immediate certainty as we have of the states of consciousness which we consider to be their effects. They are not
immediately observed facts, but results of the application
of the law of causation to those facts. Strictly speaking,
the existence of a " self " and of a " not-self " are hypotheses by which we account for the facts of consciousness.
They stand upon the same footing as the belief in the
general trustworthiness of memory, and in the general
constancy of the order of nature — as hypothetical
assumptions which cannot be proved, or known with
that highest degree of certainty which is given by immediate consciousness ; but which, nevertheless, are of
the highest practical value, inasmuch as the conclusions logically drawn from them are always verified
by experience.

This, in my judgment, is the ultimate issue of Descartes'
argument ; but it is proper for me to point out that we

have left Descartes himself some way behind us. He stopped at the famous formula, " I think, therefore I am." But a little consideration will show this formula to be full of snares and verbal entanglements. In the first place, the "therefore" has no business there. The " I am" is assumed in the " I think," which is simply another way of saying "I am thinking." And, in the second place, " I think " is not one simple proposition, but three distinct assertions rolled into one. The first of these is, "something called I exists ; " the second is, " something called thought exists;" and the third is, " the thought is the result of the action of the I."

Now, it will be obvious to you, that the only one of these three propositions which can stand the Cartesian test of certainty is the second. It cannot be doubted, for the very doubt is an existent thought. But the first and third, whether true or not, may be doubted, and have been doubted. For the assertor may be asked, How do you know that thought is not self-existent ; or that a given thought is not the effect of its antecedent thought, or of some external power ? And a diversity of other questions, much more easily put than answered. Descartes, determined as he was to strip off all the garments which the intellect weaves for itself, forgot this gossamer shirt of the "self;" to the great detriment, and indeed ruin, of his toilet when he began to clothe himself again.

But it is beside my purpose to dwell upon the minor peculiarities of the Cartesian philosophy. All I wish to put clearly before your minds thus far, is that Descartes, having commenced by declaring doubt to be a duty, found certainty in consciousness alone ; and that the necessary outcome of his views is what may properly be termed Idealism; namely, the doctrine that, whatever the universe may be, all we can know of it is the picture presented to us by consciousness. This picture may be a true likeness—though how this can be is inconceivable; or it may have no more resemblance to its cause than one of Bach's fugues has to the person who is playing it; or than a piece of poetry has to the mouth and lips of a reciter. It is enough for all the practical purposes of human existence if we find that our trust in the representations of consciousness is verified by results; and that, by their help, we are enabled "to walk sure-footedly in this life."

Thus the method, or path which leads to truth, indicated by Descartes, takes us straight to the Critical Idealism of his great successor Kant. It is that Idealism which declares the ultimate fact of all knowledge to be a consciousness, or, in other words, a mental phænomenon; and therefore affirms the highest of all certainties, and indeed the only absolute certainty, to be the existence of mind. But it is also that Idealism which refuses to make any assertions, either positive or negative, as to what lies beyond consciousness. It accuses the subtle Berkeley of stepping beyond the limits of knowledge when he declared that a substance of matter does not exist; and of illogicality, for not seeing that the arguments which he supposed demolished the existence of matter were equally destructive to the existence of soul. And it refuses to listen to the jargon of more recent days about the "Absolute," and all the other hypostatized adjectives, the initial letters of the names of which are generally printed in capital letters; just as you give a Grenadier a bearskin cap, to make him look more formidable than he is by nature.

I repeat, the path indicated and followed by Descartes which we have hitherto been treading, leads through doubt to that critical Idealism which lies at the heart of modern metaphysical thought. But the "Discourse" shows us another, and apparently very different, path, which leads, quite as definitely, to that correlation of all the phænomena of the universe with matter and motion, which lies at the heart of modern physical thought, and which most people call Materialism.

The early part of the seventeenth century, when Descartes reached manhood, is one of the great epochs of the intellectual life of mankind. At that time, physical science suddenly strode into the arena of public and familiar thought, and openly challenged, not only Philosophy and the Church, but that common ignorance which passes by the name of Common Sense. The assertion of the motion of the earth was a defiance to all three, and Physical Science threw down her glove by the hand of Galileo.

It is not pleasant to think of the immediate result of the combat; to see the champion of science, old, worn, and on his knees before the Cardinal Inquisitor, signing his name to what he knew to be a lie. And, no doubt, the Cardinals rubbed their hands as they thought how well they had silenced and discredited their adversary. But two hundred years have passed, and however feeble or faulty her soldiers, Physical Science sits crowned and enthroned as one of the legitimate rulers of the world of thought. Charity children would be ashamed not to know that the earth moves ; while the Schoolmen are forgotten ; and the Cardinals,—well, the Cardinals are at the Œcumenical Council, still at their old business of trying to stop the movement of the world.

As a ship, which having lain becalmed with every stitch of canvas set, bounds away before the breeze which springs up astern, so the mind of Descartes, poised in equilibrium of doubt, not only yielded to the full force of the impulse towards physical science and physical ways of thought, given by his great contemporaries, Galileo and Harvey, but shot beyond them ; and anticipated, by bold speculation, the conclusions, which could only be placed upon a secure foundation by the labours of generations of workers.

Descartes saw that the discoveries of Galileo meant that the remotest parts of the universe were governed by mechanical laws ; while those of Harvey meant that the same laws presided over the operations of that portion of the world which is nearest to us, namely, our own bodily frame.   And crossing the interval between the centre and its vast circumference by one of the great strides of genius, Descartes sought to resolve all the phænomena of the universe into matter and motion, or forces operating according to law.[1] This grand conception, which is sketched in the "Discours," and more fully developed in the "Principes" and in the "Traité de l'Homme," he worked out with extraordinary power and knowledge ; and with the effect of arriving, in the last-named essay,

---

[1] " Au milieu de toutes ses erreurs, il ne faut pas méconnaître une grande idée, qui consiste à avoir tenté pour la première fois de ramener tous les phénomènes naturels à n'être qu'un simple dévelloppement des lois de la mécanique," is the weighty judgment of Biot, cited by Bouillier (*Histoire de la Philosophie Cartésienne*, t. i. p. 196).

at that purely mechanical view of vital phænomena towards which modern physiology is striving.

Let us try to understand how Descartes got into this path, and why it led him where it did. The mechanism of the circulation of the blood had evidently taken a great hold of his mind, as he describes it several times, at much length. After giving a full account of it in the "Discourse," and erroneously describing the motion of the blood, not to the contraction of the walls of the heart, but to the heat which he supposes to be generated there, he adds :—

"This motion, which I have just explained, is as much the necessary result of the structure of the parts which one can see in the heart, and of the heat which one may feel there with one's fingers, and of the nature of the blood, which may be experimentally ascertained; as is that of a clock of the force, the situation, and the figure, of its weight and of its wheels."

But if this apparently vital operation were explicable as a simple mechanism, might not other vital operations be reducible to the same category? Descartes replies without hesitation in the affirmative.

"The animal spirits," says he, "resemble a very subtle fluid, or a very pure and vivid flame, and are continually generated in the heart, and ascend to the brain as to a sort of reservoir. Hence they pass into the nerves and are distributed to the muscles, causing contraction, or relaxation, according to their quantity."

Thus, according to Descartes, the animal body is an automaton, which is competent to perform all the animal functions in exactly the same way as a clock or any other piece of mechanism. As he puts the case himself :—

"In proportion as these spirits [the animal spirits] enter the cavities of the brain, they pass thence into the pores of its substance, and from these pores into the nerves; where, according as they enter, or even only tend to enter, more or less, into one than into another, they have the power of altering the figure of the muscles into which the nerves are inserted, and by this means of causing all the limbs to move. Thus, as you may have seen in the grottoes and the fountains in royal gardens, the force with which the water issues from its reservoir is sufficient to move various machines, and even to make them play instruments, or pronounce words according to the different disposition of the pipes which lead the water.

"And, in truth, the nerves of the machine which I am describing may very well be compared to the pipes of these waterworks ; its muscles and its tendons to the other various engines and springs which seem to move them ; its animal spirits to the water which impels them, of which the heart is the fountain ; while the cavities of the brain are the central office. Moreover, respiration and other such actions as are natural and usual in the body, and which depend on the course of the spirits, are like the movements of a clock, or of a mill, which may be kept up by the ordinary flow of the water.

" The external objects which, by their mere presence, act upon the organs of the senses ; and which, by this means, determine the corporal machine to move in many different ways, according as the parts of the brain are arranged, are like the strangers who, entering into some of the grottoes of these waterworks, unconsciously cause the movements which take place in their presence. For they cannot enter without treading upon certain planks so arranged that, for example, if they approach a bathing Diana, they cause her to hide among the reeds ; and if they attempt to follow her, they see approaching a Neptune, who threatens them with his trident ; or if they try some other way, they cause some monster who vomits water into their faces, to dart out ; or like contrivances, according to the fancy of the engineers who have made them. And lastly, when the *rational soul* is lodged in this machine, it will have its principal seat in the brain, and will take the place of the engineer, who ought to be in that part of the works with which all the pipes are connected, when he wishes to increase, or to slacken, or in some way to alter, their movements." [1]

And again still more strongly :—

" All the functions which I have attributed to this machine (the body), as the digestion of food, the pulsation of the heart and of the arteries ; the nutrition and the growth of the limbs ; respiration,

---

[1] "Traité de l'Homme " (Cousin's Edition), p. 347.

wakefulness, and sleep ; the reception of light, sounds, odours, flavours, heat, and such like qualities, in the organs of the external senses ; the impression of the ideas of these in the organ of common sense and in the imagination ; the retention, or the impression, of these ideas on the memory ; the internal movements of the appetites and the passions ; and lastly, the external movements of all the limbs, which follow so aptly, as well the action of the objects which are presented to the senses, as the impressions which meet in the memory, that they imitate as nearly as possible those of a real man :[1] I desire, I say, that you should consider that these functions in the machine naturally proceed from the mere arrangement of its organs, neither more nor less than do the movements of a clock, or other automaton, from that of its weights and its wheels ; so that, so far as these are concerned, it is not necessary to conceive any other vegetative or sensitive soul, nor any other principle of motion, or of life, than the blood and the spirits agitated by the fire which burns continually in the heart, and which is no wise essentially different from all the fires which exist in inanimate bodies."[2]

The spirit of these passages is exactly that of the most advanced physiology of the present day ; all that is necessary to make them coincide with our present physiology in form, is to represent the details of the working of the animal machinery in modern language, and by the aid of modern conceptions.

Most undoubtedly, the digestion of food in the human body is a purely chemical process ; and the passage of the nutritive parts of that food into the blood, a physical operation. Beyond all question, the circulation of the blood is simply a matter of mechanism, and results from the structure and arrangement of the parts of the heart and vessels, from the contractility of those organs, and

[1] Descartes pretends that he does not apply his views to the human body, but only to an imaginary machine which, if it could be constructed, would do all that the human body does ; throwing a sop to Cerberus unworthily ; and uselessly, because Cerberus was by no means stupid enough to swallow it.

[2] "Traité de l'Homme," p. 427.

from the regulation of that contractility by an automa-
tically acting nervous apparatus.   The progress of physiology has further shown, that the contractility of the
muscles and the irritability of the nerves are purely the
results of the molecular mechanism of those organs ; and
that the regular movements of the respiratory, ali-
mentary, and other internal organs are governed and
guided, as mechanically, by their appropriate nervous
centres.   The even rhythm of the breathing of every one
of us depends upon the structural integrity of a particular
region of the medulla oblongata, as much as the ticking
of a clock depends upon the integrity of the escapement.
You may take away the hands of a clock and break up its
striking machinery, but it will still tick ; and a man may
be unable to feel, speak, or move, and yet he will breathe.

Again, in entire accordance with Descartes' affirmation,
it is certain that the modes of motion which constitute
the physical basis of light, sound, and heat, are trans-
muted into affections of nervous matter by the sensory
organs.   These affections are, so to speak, a kind of
physical ideas, which are retained in the central organs,
constituting what might be called physical memory, and
may be combined in a manner which answers to association and imagination, or may give rise to muscular
contractions, in those "reflex actions" which are the
mechanical representatives of volitions.

Consider what happens when a blow is aimed at the
eye.[1]   Instantly, and without our knowledge or will, and
even against the will, the eyelids close.   What is it that
happens ?   A picture of the rapidly advancing fist is
made upon the retina at the back of the eye.   The retina

---

[1] Compare "Traité des Passions," Art. XIII. and XVI.

changes this picture into an affection of a number of the
fibres of the optic nerve ; the fibres of the optic nerve
affect certain parts of the brain ; the brain, in consequence,
affects those particular fibres of the seventh nerve which
go to the orbicular muscle of the eyelids ; the change in
these nerve-fibres causes the muscular fibres to change
their dimensions, so as to become shorter and broader ;
and the result is the closing of the slit between the two
lids, round which these fibres are disposed.   Here is a
pure mechanism, giving rise to a purposive action, and
strictly comparable to that by which Descartes supposes
his waterwork Diana to be moved.   But we may go
further, and inquire whether our volition, in what we term
voluntary action, ever plays any other part than that of
Descartes' engineer, sitting in his office, and turning this
tap or the other, as he wishes to set one or another
machine in motion, but exercising no direct influence
upon the movements of the whole.

Our voluntary acts consist of two parts : firstly, we
desire to perform a certain action ; and, secondly, we some-
how set a-going a machinery which does what we desire.
But so little do we directly influence that machinery,
that nine-tenths of us do not even know its existence.

Suppose one wills to raise one's arm and whirl it round.
Nothing is easier.   But the majority of us do not know
that nerves and muscles are concerned in this process ;
and the best anatomist among us would be amazingly
perplexed, if he were called upon to direct the succession,
and the relative strength, of the multitudinous nerve-
changes, which are the actual causes of this very simple
operation.

So again in speaking.   How many of us know that the

voice is produced in the larynx, and modified by the
mouth ? How many among these instructed persons
understand how the voice is produced and modified?
And what living man, if he had unlimited control over all
the nerves supplying the mouth and larynx of another
person, could make him pronounce a sentence ? Yet, if
one has anything to say, what is easier than to say it ?
We desire the utterance of certain words : we touch the
spring of the word-machine, and they are spoken. Just
as Descartes' engineer, when he wanted a particular hy-
draulic machine to play, had only to turn a tap, and what
he wished was done. It is because the body is a machine that education is possible. Education is the formation of habits, a superinducing of an artificial organization
upon the natural organization of the body ; so that acts,
which at first required a conscious effort, eventually
became unconscious and mechanical. If the act which
primarily requires a distinct consciousness and volition
of its details, always needed the same effort, education
would be an impossibility.

According to Descartes, then, all the functions which
are common to man and animals are performed by the
body as a mere mechanism, and he looks upon consciousness as the peculiar distinction of the " *chose pensante,*"
of the " rational soul," which in man (and in man
only, in Descartes' opinion) is superadded to the body.
This rational soul he conceived to be lodged in the
pineal gland, as in a sort of central office ; and, here, by
the intermediation of the animal spirits, it became aware
of what was going on in the body, or influenced the
operations of the body. Modern physiologists do not

ascribe so exalted a function to the little pineal gland, but, in a vague sort of way, they adopt Descartes' principle, and suppose that the soul is lodged in the cortical part of the brain—at least this is commonly regarded as the seat and instrument of consciousness.

Descartes has clearly stated what he conceived to be the difference between spirit and matter. Matter is substance which has extension, but does not think; spirit is substance which thinks, but has no extension. It is very hard to form a definite notion of what this phraseology means, when it is taken in connexion with the location of the soul in the pineal gland; and I can only represent it to myself as signifying that the soul is a mathematical point, having place but not extension, within the limits of the pineal gland. Not only has it place, but it must exert force; for, according to the hypothesis, it is competent, when it wills, to change the course of the animal spirits, which consist of matter in motion. Thus the soul becomes a centre of force. But, at the same time, the distinction between spirit and matter vanishes; inasmuch as matter, according to a tenable hypothesis, may be nothing but a multitude of centres of force. The case is worse if we adopt the modern vague notion that consciousness is seated in the grey matter of the cerebrum, generally; for, as the grey matter has extension, that which is lodged in it must also have extension. And thus we are led, in another way, to lose spirit in matter.

In truth, Descartes' physiology, like the modern physiology of which it anticipates the spirit, leads straight to Materialism, so far as that title is rightly applicable to the

doctrine that we have no knowledge of any thinking substance, apart from extended substance ; and that thought
is as much a function of matter as motion is.   Thus we
arrive at the singular result that, of the two paths opened
up to us in the " Discourse upon Method," the one
leads, by way of Berkeley and Hume, to Kant and
Idealism ; while the other leads, by way of De La
Mettrie and Priestley, to modern physiology and Materialism.[1]   Our stem divides into two main branches,
which grow in opposite ways, and bear flowers which
look as different as they can well be.   But each branch
is sound and healthy, and has as much life and vigour
as the other.

If a botanist found this state of things in a new plant,
I imagine that he might be inclined to think that his tree
was monœcious—that the flowers were of different sexes,
and that, so far from setting up a barrier between the
two branches of the tree, the only hope of fertility lay in
bringing them together.   I may be taking too much of a
naturalist's view of the case, but I must confess that this
is exactly my notion of what is to be done with metaphysics and physics.   Their differences are comple-
mentary, not antagonistic ; and thought will never be
completely fruitful until the one unites with the other.
Let me try to explain what I mean.   I hold, with the
Materialist, that the human body, like all living bodies,

---

[1] Bouillier, into whose excellent "History of the Cartesian Philosophy "
I had not looked when this passage was written, says, very justly, thát Descartes
" a merité le titre de père de la physique, aussi bien que celui de père de la
métaphysique moderne " (t. i. p. 197).   See also Kuno Fischer's "Geschichte
der neuen Philosophie," Bd. i. ; and the very remarkable work of Lange,
"Geschichte des Materialismus."—A good translation of the latter would be
a great service to philosophy in England.

is a machine, all the operations of which will, sooner or later, be explained on physical principles. I believe that we shall, sooner or later, arrive at a mechanical equivalent of consciousness, just as we have arrived at a mechanical equivalent of heat. If a pound weight falling through a distance of a foot gives rise to a definite amount of heat, which may properly be said to be its equivalent; the same pound weight falling through a foot on a man's hand gives rise to a definite amount of feeling, which might with equal propriety be said to be its equivalent in consciousness.[1] And as we already know that there is a certain parity between the intensity of a pain and the strength of one's desire to get rid of that pain; and secondly, that there is a certain correspondence between the intensity of the heat, or mechanical violence, which gives rise to the pain, and the pain itself; the possibility of the establishment of a correlation between mechanical force and volition becomes apparent. And the same conclusion is suggested by the fact that, within certain limits, the intensity of the mechanical force we exert is proportioned to the intensity of our desire to exert it.

Thus I am prepared to go with the Materialists wherever the true pursuit of the path of Descartes may lead them; and I am glad, on all occasions, to declare my belief that their fearless development of the materialistic aspect of these matters has had an immense, and a most beneficial, influence upon physiology and psychology. Nay more, when they go farther than I think they are

---

[1] For all the qualifications which need to be made here, I refer the reader to the thorough discussion of the nature of the relation between nerve-action and consciousness in Mr. Herbert Spencer's "Principles of Psychology," p. 115 *et seq.*

entitled to do—when they introduce Calvinism into science and declare that man is nothing but a machine, I do not see any particular harm in their doctrines, so long as they admit that which is a matter of experimental fact—namely, that it is a machine capable of adjusting itself within certain limits.

I protest that if some great Power would agree to make me always think what is true and do what is right, on condition of being turned into a sort of clock and wound up every morning before I got out of bed, I should instantly close with the offer. The only freedom I care about is the freedom to do right; the freedom to do wrong I am ready to part with on the cheapest terms to any one who will take it of me. But when the Materialists stray beyond the borders of their path and begin to talk about there being nothing else in the universe but Matter and Force and Necessary Laws, and all the rest of *their* "grenadiers," I decline to follow them. I go back to the point from which we started, and to the other path of Descartes. I remind you that we have already seen clearly and distinctly, and in a manner which admits of no doubt, that all our knowledge is a knowledge of states of consciousness. "Matter" and "Force" are, so far as we can know, mere names for certain forms of consciousness. "Necessary" means that of which we cannot conceive the contrary. "Law" means a rule which we have always found to hold good, and which we expect always will hold good. Thus it is an indisputable truth that what we call the material world is only known to us under the forms of the ideal world; and, as Descartes tells us, our knowledge of the soul is more intimate and certain than our knowledge of the body. If I say that impenetrability is a property of matter, all that I can really mean is that the consciousness I call extension, and the consciousness I call resistance, constantly accompany one another. Why and how they are thus related is a mystery. And if I say that thought is a property of matter, all that I can mean is that, actually or possibly, the consciousness of extension and that of resistance accompany all other sorts of consciousness. But, as in the former case, why they are thus associated is an insoluble mystery.

From all this it follows that what I may term legitimate materialism, that is, the extension of the conceptions and of the methods of physical science to the highest as well as the lowest phenomena of vitality, is neither more nor less than a sort of shorthand Idealism; and Descartes' two paths meet at the summit of the mountain, though they set out on opposite sides of it.

The reconciliation of physics and metaphysics lies in the acknowledgment of faults upon both sides; in the confession by physics that all the phænomena of nature are, in their ultimate analysis, known to us only as facts of consciousness; in the admission by metaphysics, that the facts of consciousness are, practically, interpretable only by the methods and the formulæ of physics: and, finally, in the observance by both metaphysical and physical thinkers of Descartes' maxim—assent to no proposition the matter of which is not so clear and distinct that it cannot be doubted.

When you did me the honour to ask me to deliver this

address, I confess I was perplexed what topic to select. For you are emphatically and distinctly a *Christian* body ; while science and philosophy, within the range of which lie all the topics on which I could venture to speak, are neither Christian, nor Unchristian, but are Extrachristian, and have a world of their own, which, to use language which will be very familiar to your ears just now, is not only "unsectarian," but is altogether "secular." The arguments which I have put before you to-night, for example, are not inconsistent, so far as I know, with any form of theology.

After much consideration, I thought that I might be most useful to you, if I attempted to give you some vision of this Extrachristian world, as it appears to a person who' lives a good deal in it ; and if I tried to show you by what methods the dwellers therein try to distinguish truth from falsehood, in regard to some of the deepest and most difficult problems that beset humanity, "in order to be clear about their actions, and to walk sure-footedly in this life," as Descartes says.

It struck me that if the execution of my project came anywhere near the conception of it, you would become aware that the philosophers and the men of science are not exactly what they are sometimes represented to you to be ; and that their methods and paths do not lead so perpendicularly downwards as you are occasionally told they do.   And I must admit, also, that a particular and personal motive weighed with me,—namely, the desire to show that a certain discourse, which brought a great storm about my head some time ago, contained nothing but the ultimate development of the views of the father of modern philosophy. I do not know if I have been quite wise in allowing this last motive to weigh with me. They say that the most dangerous thing one can do in a thunderstorm is to shelter oneself under a great tree, and the history of Descartes' life shows how narrowly he escaped being riven by the lightnings, which were more destructive in his time than in ours.

Descartes lived and died a good Catholic, and prided himself upon having demonstrated the existence of God and of the soul of man. As a reward for his exertions, his old friends the Jesuits put his works upon the "Index," and called him an Atheist; while the Protestant divines of Holland declared him to be both a Jesuit and an Atheist. His books narrowly escaped being burned by the hangman; the fate of Vanini was dangled before his eyes; and the misfortunes of Galileo so alarmed him, that he well-nigh renounced the pursuits by which the world has so greatly benefited, and was driven into subterfuges and evasions which were not worthy of him.

"Very cowardly," you may say; and so it was. But you must make allowance for the fact that, in the seventeenth century, not only did heresy mean possible burning, or imprisonment, but the very suspicion of it destroyed a man's peace, and rendered the calm pursuit of truth difficult or impossible. I fancy that Descartes was a man to care more about being worried and disturbed, than about being burned outright; and, like many other men, sacrificed for the sake of peace and quietness, what he would have stubbornly maintained against downright violence.

However this may be, let those who are sure they would have done better throw stones at him. I have no feelings but those of gratitude and reverence for the man who did what he did, when he did; and a sort of shame that any one should repine against taking a fair share of such treatment as the world thought good enough for him.

Finally, it occurs to me that, such being my feeling about the matter, it may be useful to all of us if I ask you, "What is yours? Do you think that the Christianity of the seventeenth century looks nobler and more attractive for such treatment of such a man?" You will hardly reply that it does. But if it does not, may it not be well if all of you do what lies within your power to prevent the Christianity of the nineteenth century from repeating the scandal?

There are one or two living men, who, a couple of centuries hence, will be remembered as Descartes is now, because they have produced great thoughts which will live and grow as long as mankind lasts.

If the twenty-first century studies their history, it will find that the Christianity of the middle of the nineteenth century recognised them only as objects of vilification. It is for you and such as you, Christian young men, to say whether this shall be as true of the Christianity of the future as it is of that of the present. I appeal to you to say " No," in your own interest, and in that of the Christianity you profess.

In the interest of Science, no appeal is needful; as Dante sings of Fortune—

> " Quest' è colei, ch'è tanto posta in croce
> Pur da color, che le dovrian dar lode

C C

> Dandole biasmo a torto e mala voce.
> Ma ella s' è beata, e ciò non ode :
> Con l' altre prime creature lieta
> Volve sua spera, e beata si gode : "[1]

so, whatever evil voices may rage, Science, secure among the powers that are eternal, will do her work and be blessed.

----

[1] " And this is she who's put on cross so much,
Even by them who ought to give her praise,
Giving her wrongly ill repute and blame.
But she is blessed, and she hears not this :
She, with the other primal creatures, glad
Revolves her sphere, and blessed joys herself."

Inferno, vii. 90—95 (W. M. Rossetti's Translation).

THE END.

Printed in the United States
By Bookmasters